Thomas Edison

The Inspirational Life Story of Thomas Edison

(Life Lessons From Thomas Edison You Never Knew Before)

Helen Boone

Published By **Elena Holly**

Helen Boone

All Rights Reserved

Thomas Edison: The Inspirational Life Story of Thomas Edison (Life Lessons From Thomas Edison You Never Knew Before)

ISBN 978-1-77485-707-6

No part of this guidebook shall be reproduced in any form without permission in writing from the publisher except in the case of brief quotations embodied in critical articles or reviews.

Legal & Disclaimer

The information contained in this ebook is not designed to replace or take the place of any form of medicine or professional medical advice. The information in this ebook has been provided for educational & entertainment purposes only.

The information contained in this book has been compiled from sources deemed reliable, and it is accurate to the best of the Author's knowledge; however, the Author cannot guarantee its accuracy and validity and cannot be held liable for any errors or omissions. Changes are periodically made to this book. You must consult your doctor or get professional medical advice before using any of the suggested remedies, techniques, or information in this book.

Upon using the information contained in this book, you agree to hold harmless the Author from and against any damages, costs, and expenses, including any legal fees potentially resulting from the application of any of the information provided by this guide. This disclaimer applies to any damages or injury caused by the use and application, whether directly or indirectly, of any advice or information presented, whether for breach of contract, tort, negligence, personal injury, criminal intent, or under any other cause of action.

You agree to accept all risks of using the information presented inside this book. You need to consult a professional medical practitioner in order to ensure you are both able and healthy enough to participate in this program.

TABLE OF CONTENTS

Introduction 1

Chapter 1: The Childhood And Deafness 6

Chapter 2: Redefining The Face Of Inventing 18

Chapter 3: Light Bulb And Notable Inventions... 28

Chapter 4: Motion Pictures And The Kinescope 45

Chapter 5: Edison's Little Well-Known Inventions 50

Chapter 6: Businessman & Industrialist............. 55

Chapter 7: Navy Career 62

Chapter 8: Last Years 68

Chapter 9: Edison's Reborn 72

Chapter 10: Edison At War............................. 102

Chapter 11: The Edison Name........................ 133

Chapter 12: Edison In Manhattan 155

Conclusion 180

Introduction

Thomas Edison, the Wizard of Menlo Park was the one who brought the United States out of the gaslight era and brought the nation into the electronic age. Since when he was just a child, Edison became fascinated with the way things operated, particularly electricity. Although he did not have formal training, Edison developed inventions and innovations that are still being used today. Edison invented innovative ways to understand and utilize electricity. The laboratories that he created and designed were by themselves innovations. He was fully equipped and the labs with staff with 1,093 patentable inventions and inventions emerged of these laboratories.

In the early 1870s , electric power and light were only a pipe dream. Edison's curiosity about technology that led him to the development of incandescent lighting which came on the market during autumn of 1878. He was determined to create an underground method of installing electric lights in private residences. Edison believed that electric light could be a safer alternative to gas-burning lamps which are in use today. Numerous lives could be saved. Edison and his ideas brought the dreams of his generation to reality.

His most famous inventions include the telephone and his incandescent light bulb the moving picture camera , and batteries. Other, less significant inventions include cement and the cement mixer and a carbon button to the phone, cement houses as well as the tattoo pen. Thomas Alva Edison was the industrial revolution's most significant innovator and inventor. His inventions had an immediate impact on America's transformation into an industrialized economy.

Edison began to work with electronics as a young man. He continued to study inventing and innovating up to the point of his death. Edison was praised for his determination and focus in the field of inventions and breakthroughs. However, Edison had his faults. There are still debates that continue to be discussed about the way Edison did not actually invent anything his ideas, but he paid large amounts of money to other people to come up with ideas for his benefit. Edison gave them ideas, then he took their ideas, patents and promoted the inventions that interested Edison. He also provided the ideas that evolved into inventions.

It is an attack on Edison's character to say that he was a marketer as well as an industrialist instead of an inventor. But, if you look at it this way there is no way to be one with the other. If you create all the inventions you can think of, but do not bring your invention to market or does not benefit others, then you've not succeeded in your endeavors. Edison certainly did not fail.

Of the 1,093 patents granted by Edison The top patents included the electric light bulb and bulbs, the lightbulb and the motion-picture camera generators, batteries, and his labs. Three of his greatest inventions and innovations led the development of new businesses: electricity utilities record and phonograph companies as well as films. In the 1920's these industries contributed an impressive 7 percent of the nation's gross production. If you read through the history books, you'll see that the time from 1879 until 1900 is called"the Age of Edison. This was the time during which Edison developed his ideas and ingenuity.

He was the inventor of many devices that have significantly changed the way we live. He was among the first innovators to apply the concepts that of large-scale production as well as mass teamwork in the process of inventing. The fourth-highest number of inventors in the history of science.

You should admire Thomas Alva Edison for his never-say-die attitude. Failures, for Edison were speed bumps on the way to success. Edison stated, "If I find 10,000 ways to fail I've not been unsuccessful. I'm not discouraged since every failed attempt is a step in the right direction," (Baskerville 2012).

Thomas A. Edison was the most responsible person in bringing us the world of today. There was no one who did more to influence the way we live today than Edison. Edison was the most influential person in his day. He was a smart businessman who knew how to sell. He would start a business and then hire scientists and engineers to help him develop his ideas and create an end product. He enjoyed many success stories however, he also had a few failures however, he continued to work. He continues to be an inspiration to entrepreneurs today.

It is said that Edison was the product of his time. The latter part of the 1800s and the beginning of 1900s was the time of the Industrial Revolution in the United States.

There were many scientific breakthroughs making progress during this period and Edison was among America's top industrialists. His ventures were not all had a great success. He made a number of mistakes and was ecstatic to take on another project. One time, when a person in his lab expressed discontent over the slow progress made on the battery for storage, Edison exclaimed, ""Why you ask me, I've got lots of results. I've got a lot of items that aren't working!" (Vernon, 2014).

Chapter 1: The Childhood And Deafness

"Your worth is in the things you're worth, not what you own." * T. Edison

"The Wizard of Menlo Park," Thomas Alva Edison was born on the 11th of February 1847. Edison was an outstanding American inventor and entrepreneur, as well as a marketer. Without Edison the world could be much less prosperous. There may not be a motion picture camera, phonograph or even an electric light bulb. Edison applied the principles of teamwork and mass production to the processes of his invention. Furthermore, Edison created the first industrial research lab. In this article, we'll examine his life in the early years and what led him to the path of ingenuity!

Childhood

Edison began his life in Milan, Ohio. He was born into an upper middle class family with a bustling river port bustle, and the activity eventually inspired Edison's genius in the development of new technology. In the later years of his life, he set up a laboratory located in Menlo Park, New Jersey and invented the telegraph. He also created the first electric lightbulb the phone, the alkaline storage batteries as well as cameras for motion pictures. He dispatched an assistant to an isolated town in Idaho to demonstrate the use on his Kinetograph. A young boy on the journey to school stopped to look at the amazing invention of Edison. He observed as Edison's assistant flicked over his Kinetograph and animated images moved across the screen. "It is very pleasant" was what Edison would tell his family after a while.

Edison was the eldest of seven children born to Samuel Edison and Nancy Edison. His father was a exiled political militant who was exiled from Canada while his mother was was a teacher in a school. When Thomas was a young boy the teen suffered from scarlet fever and severe ear infections. The ear infections caused him to have hearing problems for both ears. Hearing problems remained as he got older and eventually led him to become nearly deaf in the latter years of his life. Edison was always of the opinion that his hearing loss was caused by an incident on the train that damaged his ears.

Edison was not able to speak until almost 4 years of age. But, once he started to speak and talk, he would ask his companions to explain the way in which everything worked.

If they claimed that they didn't have any idea, Edison would look at them and ask "Why?" (Beal, 1999).

Edison had a swollen kid Edison was ill, but he also extremely interested. Edison once asked his mom what the reason why geese sat upon their eggs. Nancy Edison explained to Thomas that they do this to make eggs. Then, a couple of hours later, when Thomas was missing, she discovered him sitting in a nest of eggs from geese.

Edison's father was extremely strict, even to the level of abuse. When Thomas was able to burn down the barn of the family aged six, the boy was publicly punished following an announcement to the local community. While this made an irresistible impression on the boy Thomas, he continued his inquiring and sometimes bizarre manners (Vernon 2014).).

Training and the First Job

Edison relocated from Port Huron, Michigan in 1854. He went to the local public school for a period of 12 weeks. He was extremely hyperactive with a tendency to be distracted. He was classified as a difficult pupil. The teacher believed his bigger than average head and wide forehead was a sign of an overactive brain. She concluded that Edison could not learn. The teacher didn't know how to deal with a child who was always asking questions and challenging her responses. Edison's mother, furious at the move, removed Edison from the school and instructed him to instruct Thomas at the home. "[Shewas] the creator of my character... because she always was so honest and so confident in myself ... she always gave me the impression that that I had someone worth living for and I was not able to disappoint her," (Edison, 1854).

Nancy Edison, Thomas' mother, was a fervent and beautiful daughter of an Presbyterian minister and an outstanding educator. She taught her child to master the "Three R's" (reading writing, reading, and math) as well as the Bible. His somewhat shrewd father, politician Samuel Edison encouraged Thomas to read the best classics. For each classic Thomas would finish his father offered him the sum of ten cents to reward him. Samuel Edison introduced the young Thomas Edison to Thomas Paine who was a radical philosopher and political activist during his time during the Revolutionary War period. Reading classics and philosophical texts quickly became Edison's love affair. Edison particularly enjoyed popular science publications and novels. He devoured books and read constantly and continued to read throughout his life.

Thomas Edison developed a highly individualistic style of learning. He challenged the accepted theories about electricity, and approached this nebulous subject with a view to reinventing the theories of his time. He utilized his skills, memory, and perseverance to carry out experiments that validated his theories. He refined his style of analysis with a bold stance by first doubting the method and then rigorously trying it out for himself. He was always interested keen, competitive, diligent, and extremely hardworking.

Young Edison was a fan of writing poetry and reading it. He would often recite to Gray's Elegy in the Country Churchyard, "The boast of heraldry that oozes the power and pomp and all the beauty the wealth you gave and all of us await the inevitable time. The path to glory takes you to the grave,"(Beals 1999).

When Edison was twelve years old, Edison took the initiative of his age to put his comparatively small education and a shrewd mind to use. Edison started selling newspaper to people on the Grand Trunk Railroad Line.

This allowed him to access news bulletins that were teletyped into the train station each day. He used the teletyped news bulletins to begin creating his own publication, The Grand Trunk Herald. He also accessed to the press releases regarding the Lincoln/Douglas war debates and put them into his newspaper. His most current and amusing articles were a huge hit among train travelers. The Grand Trunk Herald was Edison's first venture into entrepreneurship. Edison saw a need and took advantage of the chance.

Thomas Edison distributed campaign literature for Abraham Lincoln and sold flattering images of the presidential candidate. The modest publishing enterprise generated an average of ten dollars each day. The amount more than paid for his expenses and he used remaining funds to build with a laboratory for chemicals located in the basement at his parents' home. His mother was upset over the smells and hazards of the chemicals that were contaminating her home. He agreed with his mother and transferred the chemicals to a locked area at the bottom of their house and the locker room of the train.

The fact that he kept his chemicals in the train proved to be dangerous. A few days ago, the train swerved across a rough stretch of track. A stick of phosphorous crashed on the floor and caught fire. A baggage vehicle caught on fire. The conductor was so furious that he verbally rebuked Edison and slammed him both sides of his head. The beating increased Thomas's already diminished hearing. Due to the fire, the stationmaster stopped Thomas from selling his newspapers on the train, even though Thomas continued to sell newspapers at the station.

There is an alternative theory to this tale that states that while Edison was getting on the train, he struggled to climb the steps of the freight car, with arms full of newspapers. He struggled to maintain his balance, and the conductor took him by the ear and threw him down. Edison noticed something crack near his ears. He declared this to be the beginning of hearing loss which got more severe.

A few years later, an operation was developed that might have restored Edison's hearing. Edison was not willing to have the procedure performed. He stated that his biggest fear was that of the possibility that he "would be unable to learn how to control his thoughts in a constantly changing world" (Beals 1999). He said that even though the last time he heard the sound of a bird since he was 12years old, the fact that he was in a better position and was able to study and read with no distractions from the conversation that swung around him.

Telegraph Operator

While hawking his paper, Edison saved a 3-year-old boy from being hit by a train that was speeding by. To show his gratitude, the dad, James MacKenzie, a Telegraph operator, instructed Edison the operation of a telegraph. Utilizing McKenzie's instruction, Edison became a part-time Telegraph operator within Port Huron, Michigan.

By the age 15 Edison was employed as a full-time Telegraph operator. For five years Edison was employed as a replacement to operators that had resigned their posts to fight during the Civil War. His reputation as an experienced operator of telegraphs gave him the chance to work with the most prestigious press-wire operators. He often sat in their offices to discuss the latest developments in technology and politics. Furthermore, Edison read widely, was a student and experimented with the telegraph system, and rapidly became an authority on the workings of the Telegraph.

When Edison first began operating the telegraph system, the messages were recorded, and it was simple for Edison to understand the patterns. With the advancement of technology but receivers came outfitted with an audio keyboard. The sound keys allowed telegraphers to "read" the messages using the sound of clicks. Due to his hearing impairment it became increasingly difficult to read the messages sent by telegraph. To make up for his loss of hearing Edison was able to develop new ideas and enhance the current Telegraph system.

In 1866, Thomas relocated into Louisville, Kentucky where he was employed by The Associated Press Bureau News wire service, which was responsible for telegraphing news stories. Thomas worked in the night shift that allowed him to read and play. While experimenting with a lead-acid battery Edison dropped sulfuric acid on the floors of his second floor room. The sulfuric acid was dripped between the floorboards and through the first floor and over the desk of his boss. The following morning, Edison was fired.

Thomas Edison returned home in 1867, at the tender age of 21 to discover that his mom was suffering from severe mental illness, possibly dementia and his father not able to work. The family was in financial hardship. Edison realized that he would get no financial support from his parents and it was his responsibility to determine his future.

He relocated to Boston and was offered a position in the Western Union Company. It was 1868 and Boston became the hub of culture and science. This was exactly what Edison required in terms of science and culture.

While he wasn't in the office of Western Union, he was creating an electronic recorder to count legislative votes. The machine was quickly patentable. Dewitt Roberts, a close acquaintance of Edison's bought an interest in the machine for the price of $100. Dewitt tried to sell the machine to Congress but Congress did not want to be involved with a device that could improve how fast people could vote. They worried that speeding up voting could reduce the time needed for the wheeling and dealing of politicians. Edison's most interesting invention was never made (Barksdale 2014).

Edison relocated into New York City in 1869 and joined forces the company with Franklin Pope and James Ashley. They created Pope, Edison and Company and claimed to be electrical engineers and builders of electrical equipment. Edison submitted and was awarded numerous patents to improve the Telegraph. The partnership was merged in 1880 together with the Gold and Stock Telegraph Company.

In the following years, Edison established the Newark Telegraph Works in Newark, New Jersey. The company's partner is William Unger and they began manufacturing printers for stock. Then, he established The American Telegraph Works and continued to work on the development of an automated Telegraph.

In just 9 months Edison was able to prove himself an important inventor for Western Union and the Gold and Stock Telegraph Company. Edison developed a more efficient stock ticker as well as his Universal Stock Printer and a sort of facsimile system which synchronized stock tickers' transactions. It was said that the Gold and Stock Telegraph Company was awestruck. They offered Edison $40,000 to acquire the rights of the machine.

Edison accepted the money, resigned from his job as a telegrapher, and started dedicating himself entirely to the creation of inventions.

Chapter 2: Redefining The Face Of Inventing

"Genius has one percent of inspiration and ninety nine percent perspiration." Thomas Alva Edison Harper's Monthly September 1932.

Edison employed a variety of machines and assistants to implement the ideas to develop his ideas. Edison also had time to establish partnerships and create telegraph products to the most expensive bidder. Western Union Telegraph Company was the primary client, but Edison's inventions were often sold to auction and he offered them at auction to bidders who could win to make a profits.

Edison was a fierce competitor. He was determined to win and patent his invention before any other person could. The majority of his inventions that were initiated and completed were promoted for the potential profits they could yield. Edison was popularly referred to as a "tight-fist" and kept his cash in his tights.

In one instance He cheated Nikola Tesla out of a guarantee of a commission when Tesla constructed and repaired numerous electricity generators that were sold to Edison. He also attempted to denigrate Tesla who he believed to be a great engineer and inventor however, he was also a rival in the battle for electrical dominance. Edison knew that if some ideas of Tesla's creations were sold as a result, they could eat the profits of his company.

One of Edison's most important ideas to help Western Union was quadruplex telegraphy. The invention was capable of sending two different signals at distinct directions over that same cable. Railroad billionaire Jay Gould found this invention intriguing and sought to take the product straight from under Western Union's eyes. Jay Gould paid Edison more than 100,000 in bonds, cash and stocks for his quadruple Telegraphy. Furthermore, Gould provided litigating representation for Edison against Western Union.

By the year 1870, Edison had set up an experimental lab and manufacturing facility located in Newark, New Jersey. Edison and his team of engineers created and manufactured an automatic Telegraph, a system of high-speed made of punched paper tape and mechanical transmitters. They were designed to compete against the manually keyed Morse Telegraphy.

Edison never was concerned about stepping onto anyone's feet. He also didn't expect his employees to be more productive than he was. He worked for long hours working into and out and inventing new items simultaneously. Edison was the most efficient multi-tasker. Employees would work for all day long with Edison because they believed in his dedication to science and his ability to apply techniques of one experiment to initiate another one.

In the 1870s Edison gained a reputation as the most innovative telegraph innovator in America. Businesses competed to control his work. In October 1870, he founded his own company, the Automatic Telegraph Company with the support of a group of wealthy financiers. The company was later renamed an organization called the American Telegraph Works, which was a massive shop with machines and skilled engineers. He was appointed the electrical consultant and mechanic for the Gold and Stock Telegraph Company while he built up his fortune, win admirationand build businesses.

Edison's wealth has brought him closer to his wife and his family. He got married at the age of 16 to Mary Stilwell in 1871.She was employed at one of his companies, and they'd known for just two months.

Menlo Park and Family

The year 1876 was a result of a suit brought by a landlord Edison transferred his inventions to the countryside of Menlo Park, New Jersey. His father was in charge of the construction of a structure to be the epitomize of all that Edison knew about the process of inventing. Edison set up his own personal research laboratory in the industrial sector and spent many moments creating. It was the first of its kind in the world. and development center.

A large portion of Edison's inventions from his were developed during his first year during his time at Menlo Park were devoted to creating systems that could be used for several telegraph systems that were used by Western Union. Because of the extensive work the Edison laboratory was doing to support Western Union, Edison proposed that Western Union support the machine shop by providing a weekly stipend of $100. The agreement was signed and provided Western Union the rights to every Edison invention in the field of telegraph.

While in his laboratory, Edison would come up with innovative ideas. He urged his assistants to collaborate with Edison and would often take credit for their efforts. Edison had a great entrepreneurial spirit. He was able to start new companies and come up with new projects. He was always convinced that there were new and exciting worlds to explore. Edison recognized the needs and set about creating it. Edison was an expert in promoting himself and his inventions in order to attract investment and financing. He was adamant about getting publicity which led to money to fund his ventures. He understood that it was important to have the media at his disposal to help further his goals.

As Edison was becoming more famous and well-known, people went in Menlo Park to see his inventions in person. Edison was always entertaining investors and businessmen who came to offer him money to fund his inventions. He filed for more than 400 patents while working at Menlo Park.

But, they often were a source of irritation for Edison. He was a cigar-smoker as well. One of his biggest pet peeves was guests and reporters who took the cigars from a box inside his office. He had tried to put his cigars in a safeplace, however, his colleagues would take the cigars out for guests. In desperate times, he devised the idea along with the cigar maker create an assortment of cigars using fillers made of old paper and hair. After a lengthy discussion with investors, Edison realized that the cigars were gone. He was quite upset after further examination, but Edison realized that he had actually consumed these "false" cigarettes himself (Vernon 2014).

Family

Thomas as well as Mary had been together for thirteen years, and they have three sons. Thomas named his first daughter Marion and affectionately named her "Dot" to refer the Morse code. The younger brother of her was named Thomas, Jr. and was referred to as "Dash."

Dash utilized his famed Edison name to market his questionable inventions and quack remedies. Thomas, Sr. requested Dash change his name to Thomas Willard to prevent harm to the Edison name. Dash's attempts to invent were utter failures, as well as his farm for mushrooms was a disaster. His inclination to invent is lost within the muck of his insanity.

Edison stated about his son "I did not manage to get him to attend school or even work within the Laboratory. Therefore, he is completely unliterate both scientifically and in other ways" (Rondeau in 2010,).

Marion Estelle Edison Oeser, Edison's very first child, spent much spending time with her dad following her mother's passing. He would often take her out to buy cigarettes for him. Marion was married to an German soldier on 1892. She resided in Germany through WWI.

The third son, William spent his time working with the Army and was attempting to make use of his Edison name to avoid being discharged. William along with his father were not friends and Edison once reacted to an inquiry for money from Williams's wife, writing "I do not see any reason to support my son. He hasn't done me any honour and has brought a gush of shame on my cheeks a number of times" (NPS 2014).

Mary Edison dearly doted on her brilliant husband and allowed him to come up with ideas. Whatever choices he made to take she was sure to be happy. Edison however, wasn't a very attentive father. He would spend his days at his laboratory inventing and disregarded his obligations towards his children. His lack of interest led to his family becoming distant from them and they never would speak highly of Edison.

Edison did not write about his personal life, however in his diary , he wrote about his wife "Mrs Mary Edison My wife Dearly Beloved is unable to invent an ounce!" And "My Wife Popsy Wopsy Can't invent!" (Santaso, 2008). His wife seldom saw his face. He would take "cat napping" all day, and he would spend all day in the lab. Edison was a hard worker who could work for up to 60 hours on his pet projects. Infrequently, Edison was working on several projects at the same time. He was not very vocal regarding his family or his wife up to her death in 1884, either from neurotumor, poisoning with morphine or typhoid. He told his friends that he was missing her deeply and showed symptoms of a depressed widower.

Edison, who was widowed for two long years was tired of living without the love of his life and trying to take care of three kids. He began to look for the right woman and was first introduced to Mina Miller by his friends the Gilliards. In the early days, Edison had a lot of money and was was becoming famous. Edison was regarded as an "catch" but because of his bug-eyes and halitosis as well as his hearing loss, Edison was considered to be a bit creepy. Edison was enthralled by Mina and she reciprocated his love. Friends told him, "Edison found his own version of paradise in Fort Myers, then a small town, and determined that he had to accomplish three things: construct an indoor winter home in Florida to marry Mina and take Mina back to the tropical Eden" (Venable 2011.).

Edison continued to instruct his spouse Mina Morse how to code. They often sat in the company of their parents and secretly write messages in each others' hands. Utilizing this talent, Edison proposed to Mina Miller by putting his proposal in her hand then she replied "yes" and they got married in 1886.

Edison did not hesitate to complain that his wife was not able to invent. But she could invest as a witness, and also consulting. Mina often kept notes for Edison while he was working, and Mina was named as a witness on some of his patents.

He bought the estate of West Orange, New Jersey as a wedding present to Mina and also to accommodate his expanding family.

Mina was Edison's PR department. She helped him maintain his privacy, while enjoying an independent life. She brought forth the first embers of an environmentalist and her efforts for conservation of nature may be the start for the creation of a national park.

Mina Thomas Thomas enjoyed a wonderful relationship. Mina was aware of Thomas's genius and decided to share his genius in the entire world. Their union was one of love, respect and goal. If Mina wasn't there to give Thomas with a safe and loving atmosphere, Thomas might not have had the success that he gained.

Thomas Edison and Mina had three children. Charles Edison who took over the business after Edison's death, Theodore Edison who had more than 80 patents on Physics and Madeleine who got married John Eyre Sloane, who established an aircraft manufacturing plant located in Long Island City in 1912.

Chapter 3: Light Bulb And Notable Inventions

"My primary goal of life is earn enough money to be able to develop more innovations." Thomas Edison

While Thomas Edison was not the person who invented electrical light bulbs but he was the one who created the first light bulb that could be sold commercially. Light bulbs from earlier times were being developed and the majority were still in development in that time. invention.

Prior to the invention of electric lighting that lasted for a long time oil lamps, or even manufactured natural gas were used to light the room. This type of lighting was responsible for numerous house fires that resulted in a number of deaths. Edison knew that something had to be done, and his experiments with electricity proved to be the solution.

In order to create a long-lasting electric light source, Edison sought financial support and hired a group of talented scientists and technicians and put his expertise to making the cheap electric lamp. Edison and his team tried many theories, in the process. Edison believed that each failure helped him get closer to achieving success.

He began his research into incandescent light bulbs in 1878, and registered his first patent later in the year. He developed and tested various metal filaments. He tried platinum and it was weakened by heat. Platinum was costly to make electricity from and was also very low in resistance. It required copper conductors to be used in the electrical distribution network to provide commercial installations of light bulbs. Later , this system utilized platinum and copper for its wiring, and was the basis for modern electricity distribution centers for utility companies.

Edison turned to carbon-based, high-resistance filament to light his bulb. In the fall of 1879, Edison tried his new filament and it was able to burn to 13.5 hours. Edison was thrilled however he continued to refine his invention and eventually filed patents for an electric lamp. "A carbon filament (or strip) of wire coiled to be joined ... via contact wires with patina" (Edison Innovation Foundations, 2011,).

After the filing and receipt of the patent issued to carbonized filaments Edison as well as his group discovered that bamboo carbonized filaments could last for more than 1200 hours. Edison quickly came up with this innovative type of filament , and then began advertising commercial lamps that used carbonized Japanese bamboo filaments. Electric lamps are now safe, reliable and affordable to purchase. Industries began to install the bulbs in increasing quantities. Edison was granted his first lighting patent for incandescent lamps on the 27th of January 1880. The patent was a step in the direction of worldwide usage in the field of electricity.

To demonstrate that light bulbs were indeed the future's light bulb He laid out an underground system that was experimental and set up lamps in Menlo Park. He was able to test the first underground electrical system in 1880.

Edison began to build an enterprise that could supply electricity to illuminate the cities around the globe. He created the Edison Illuminating Company which was the first electric utility owned by investors. Edison established a massive generator plant located on Pearl Street in New York City. Edison placed around four hundred of his incandescent lamps in commercial and residential properties. In 1882, on September

4 hundreds of New Yorkers congregated in Pearl Street. Around 3:00 pm, the generator started which caused Pearl Street lit up with electricity. This massive achievement for Edison established his theories and eventually led to the creation of a central power station. Edison had always dreamed of the possibility of a generator station to generate electricity in bulk, and now he was able to achieve it. He continued to earn fame fortune, fortune, and patents.

Businesses once again came to Menlo Park to see Edison's innovations and to put money into his inventions. However, Edison moved his business into New York City and became concentrated on manufacturing, marketing and installation. He continued to use Menlo Park, but his professional career had shifted into other locations.

Edison was the first to use incandescent lighting system used in his Paris Lighting Exhibition in 1881 as well as the Crystal Palace in London in 1882. In 1882 The Pearl Street generating station sent out 110 volts of electricity to around 60 customers in the lower Manhattan. This was the point of origin for his business. His achievements were noticed and competitors started coming out from the woodwork.

Three factors were responsible for Edison's success with incandescent light. They had discovered a strong incandescent material, removing air out of the bulb, creating the vacuum, and identifying filament materials with high resistance.

To increase the power of Edison's electric light bulb Edison invented and developed his electric generator. It was a device that could control the flow of electricity to machines. He invented the generator in 1881.

George Westinghouse, a proponent of Tesla's alternative current , or AC, was a major rival and was opposed to the direct current of Edison, also known as DC current. To increase the influence of Edison's innovations the Edison General Electric Company merged with another company in the year 1982 and changed its name to General Electric Company. Their objective was to spread DC electricity and fight Westinghouse. But, in 1889 AC electricity had begun to dominate the electric industry.

The Telephone

Western Union petitioned for Edison to design a communications device that could rival Alexander Graham Bell's phone. Bell's invention was amazing and impressive, however, it was not able to communicate voice over a considerable distance. Edison used the telephone to take another path, and as his telegraph research slows down, his research on telephones was speeding up. Edison was most interested in telegraphy, but he was familiar enough with the telephone to dig deeper into the technological advances it brought.

Edison believed that there were several ways in which Bell's telephone might be enhanced. Bell's telephone was built around the sounds of the human voice. these vibrations were converted into electrical currents, which were then heard by a different phone device. The sound was a little tinny and then became less so over longer distances. Edison discovered that he could improve the quality of his listening and speaking. He also developed an element of the telephone, which had the valve that could open or close to regulate sound and speaking.

The carbon button Transmitter was the contribution of Edison for the phone. Edison believed that one device could function as both a receiver and transmitter simultaneously. He designed a phone that included a device to be used for speaking the other, while another was specifically designed to listen (Amicevksi 2015).

William Orton, President of Western Union paid Edison $100,000 for the Carbon Button Transmitter. It was at this time that the American Speaking Telephone Company began to rival Bell and battles for legally-enforceable rights of the phone patent began to take shape.

The rivalry among Edison and Bell was over when they joined forces to create the United Telephone Company. Edison received a significant amount of money for his invention of the telephone, as well as his contribution in his work with the United Telephone Company.

The Phonograph

The phonograph was among Edison's most well-known inventions. Edison was working on improving the telegraph transmitter, when he observed that the tape in the device made the sound of speech when playing at rapid speed. He took the diaphragm from the telephone receiver and added the needle. The experiment began with a diaphragm which had an embossing ring and was held against a rapidly shifting paraffin papers. Talking caused vibrations and created marks on the paper. Edison continued to conduct this experiment believing that a stylus or needle needle could puncture paper tape and record the message. Edison then tried using a needle in an tinfoil cylinder.

The machine was designed by the inventor to include two diaphragms as well as needle units. One was used for recording, and the second one for playback. All one had to do was to speak into a mouthpiece and the vibrations would be inserted into the cylinder using the form of a groove. Edison presented his design of the machine to his mechanic John Kreusi, to build. Kreusi constructed an engine following Edison's blueprints in just 30 hours. Edison quickly had the message "Mary had a lamb" recorded. His surprise and amazement the message was replayed to Edison.

The importance of telegraph and telephone work prevented Edison's team from perfecting the phonograph over the course of five months, however in December 1877, the device was displayed in the headquarters at Scientific American. "Mr. Thomas A. Edison recently visited this office, set a tiny device on the desk and turned a knob and the machine asked us about our health condition, asked whether we were satisfied with the phonograph. It then informed us that the phonograph was working well, and then wished us a good night" (Bellis 2014). The phonograph's model recorded sounds on tin foil that was wrapped around an engraved cylinder. It was of a low audio quality it could be used only a handful of times, but it was a phenomenon which led to making Edison famous and extremely wealthy.

Alexander Graham Bell, who was the winner of the Volta Prize form the French Government and collected his winnings of $10,000 and set up a lab for acoustical research. Bell's company was working to improve Edison's phonograph design and started using wax instead of tin foil, and floating styluses instead of a needle that was rigid. Their innovations resulted in the needle being able to strike rather than indent the cylindrical. Patents were granted C. C. Bell in 1886 and the device was marketed under the name graphophone. In a meeting with Edison in order to negotiate a possible partnership with the machine, Bell believed they could improve the phonograph into an extremely marketable device. Edison declined. He was determined to develop the phonograph by himself.

Edison closely observed the improvements that were made through Bell with his phonograph, and started making use of wax-based cylinders. Edison called the phonograph he invented his "invention" his New Phonograph.

It was in 1887 that Edison created his Edison Phonograph Company with the intention of selling his brand new machine. He promoted the phonograph as an "multi-use" device. He believed that it could assist in writing letters and dictation, as well as be employed as a tool for blind people and as a recorder for the family device, as a music box and clocks that would announce the time, and even in toys. He believed that the machine could also be connected with the telephone to record telephone calls. The wax cylinders he used were white and comprised of ceresin, beeswax, as well as the stearic wax.

Jesse H. Lippincott took the control of all phonograph companies after becoming the only the licensee for the American Graphophone Company and purchasing

Edison Phonograph Company. He founded The North American Phonograph Company in 1888, and then began marketing the phonograph solely as an dictating machine for business.

Lippincott fell ill and was forced to relinquish the control over The North American Phonograph Company in 1890. It was then sold to Edison as the largest financial backer of the company. Edison did not change anything about the business, except for selling the machines instead of leasing them out to companies.

When Lippincott was busy selling their phonographs to the public, Edison's firm was making chat dolls in The Edison Phonograph Toy Manufacturing Company. They were created using tiny wax cylinders, however Edison promptly ended the company within a year. They are very rare in the present, so if you have one, you should keep it.

Then to the Edison Phonograph Company began producing musical cylinders and coin-slot phonographs were becoming more popular at dance clubs. Jukeboxes were among the first phonographs that were employed as entertainment devices (Bellis 2012).

The Spring Motor Phonograph was announced in 1895. However, Edison was unable to sell his phonographs due to legal problems surrounding the bankruptcies of the North American Phonograph Company. After a few years and a number of legal fights, Edison bought back his rights to phonographs. Edison established The National Phonograph Company. He is now able to produce Phonographs for entertainment at home.

Edison created a method that used identical wax cylinders to the phonograph, in 1901. an even harder wax was developed. This process was referred to as the Gold Molded

Cylinder due to the golden vapor released from the gold electrodes employed to manufacture. Then, between 120 and 150 Cylinders could be made every day. In 1904, the mass duplicate of these cylinders decreased prices to $0.35 per piece. The color was black with the edges being beveled.

Edison kept improving his home entertainment phonograph and, in 1909, an expensive phonograph that claimed first-class performance was launched. The machine was designed to compete with Victrola or Grafonola.

The year 1905 was the time that Edison developed the recording phonograph intended for business purposes. But, the early machines were difficult to operate and were a bit fragile. There were improvements made to Edison's business recording machine however, Columbia was still the dominant market by introducing dictaphones instead of Edison's phonograph dictation devices. In the end, Edison's company switched the electric motors on the business phonograph using spring motors. This meant that the business phonograph became capable of dictating. Edison's new and improved machine, the Ediphone was released worldwide in the year 1916, and sales increased exponentially following World War I.

Edison continued to manufacture Phonograph cylinders. In 1912, the company's main rival, Columbia, abandoned the market for cylinders. It was the United States Phonograph Company ceased production of U.S. Everlasting cylinders in 1913, and then left the recording market for cylinders to Edison.

Cylinders had risen to their peak at this point as the disk (or record) was used in the phonograph. Edison was quick to admit his disc's real and declared the creation of an Edison Disc Phonograph. However, it is interesting to note that Edison continued to produce Blue Amberol Cylinders until the company was disbanded in 1929.

Edison's research into the phonograph inspired him to develop an electric pen. The pen was patentable in 1877. Its predecessor was a pen with perforations that punched holes to create messages for telegraphs. This pen was electric and created an image of the stencil that the operator of telegraphs wrote. This stencil was used by telegraph operators to apply ink on other papers and create duplicates. The world now had one of the first machines to copy.

Alkaline Storage Batteries

One of Edison's favorite inventions was the alkaline storage device that he patent on July 1st 1906.

In the course of testing the improved version of my iron nickel batteries by using an alkaline solution it was found that active suliids existed in the battery, which reduced the effectiveness of the battery and removing from its potential as power source to be used in automobiles, where the most energy must be represented with minimal weight. To determine the source of the development the active sulids, I found that their source was caused by the existence of sulfur free within the rubber insulators, or supports (both soft and hard) which are utilized to construct the battery mechanically. This includes, for example, as separation between electrodes, as support for electrode plates. as well as the stuiiing boxes that the electrodes travel through. Naturally, any sulfur that is present could be linked to the process of vulcanizing (Edison 1904.)

Edison's aim was to develop batteries that were better-performing and less heavy than lead batteries which were popular during his time. He believed that the alkaline battery could eventually become the base of an electric car. He wanted to create batteries that were lighter than lead-acid batteries still being used in modern cars.

Edison and his team of researchers conducted tests on all kinds of metals and other substances to find pieces that can cooperate to make batteries. The idea was to create batteries that were more robust and 3 times more robust. The number of tests conducted was in the thousands and it took a long time to perfect this invention. Edison's battery eventually worked with potassium hydroxide. The component reacted to the battery's iron and nickel electrodes, generating electricity. Edison had come up with an electrical battery that was robust in its output and rechargeable.

Edison as usual manner, announced the launch of his battery in a complete ceremony. Edison made confident statements about the battery's performance and the manufacturers and owners for electric vehicles began purchasing the batteries. However, battery failures were common. Certain batteries started to leak, and others lost power after a brief trip.

Edison closed his factory in order to design his battery. He devised the new design with materials that were more expensive, however by 1910 , the battery was manufactured at Edison's New Jersey laboratory.

At the time Edison had developed his battery, the electric car was an idea from the past, given that Henry Ford had introduced the Model T car with a gasoline engine. Battery use was pushed back to special commercial automobiles. By 1912 , gasoline cars could utilize batteries to power their starters.

However, Edison's battery was not sufficiently high an i.d., and therefore was not suitable for this. The battery was reliable, however and became popular for offering backup power to railroad crossings and was utilized in mining lamps. The battery would become among Edison most lucrative money makers during his final years.

The first time the alkaline storage battery was designed to be the energy source for the phonograph as well as the electric automobile. Later, Edison invented batteries which were supplied to the U.S. Navy for submarine use. In 1912, Henry Ford, founder of the Ford Motor Company, commissioned Edison to develop the self-starter battery in Ford's Model T car. The Edison battery was never utilized, however it is interesting to consider that this initial request was just one of the many that were made by two great American entrepreneurs.

Chapter 4: Motion Pictures And The Kinescope

"If we did all the things that we're capable of, we could be awestruck .,.." Thomas Edison.

Based on his experiences with the phonograph Edison thought of the need to create "an instrument that can do what it can do for your Eye exactly what the phonograph can do for the Ear that is record and reproduce of objects moving" (Ford 2013). Edison had the concept of connecting the phonograph with the device that strung photographs. Together alongside William K.L. Dickson, Edison developed a motion picture camera which was operational, called the Kinetoscope. The first motion picture display device was made for films to be watched by one person at a time , via an open-air viewer. The Kinetoscope wasn't a film projector, in the sense of however, it did present the fundamental theory that later became the norm for cinematic projection. He patents this invention in 1891.

The original Edison device soon developed into a device made of strips of celluloid film. The film was acquired and developed with the help of George Eastman of the Eastman Kodak Company

Edison constructed a motion-picture production studio in the West Orange, New Jersey laboratory in 1893. The studio was unique because it had a roofing system that was able to be opened to let sunlight in. The entire building was set on pivots that allowed it to be moved around to ensure it was in alignment to the solar. The studio for film was referred to for its nickname, the Black Maria because it resembled an old police vehicle called"the "Black Maria."

Edison's first motion picture , and the first motion image to be copiedrighted, featured his employee, Fred Ott, pretending to cough. This invention was a huge success.

"The Sneeze" was noted because of its closeness. The next fight of James Corbett and Peter Courtney was shot on the Black Maria.

Edison employed W.K.L. Dickson to produce more than 75 film for Edison in 1894. The films were intended be aimed at male audiences and included women in sexy clothes. The performances of the strongman Eugene Sandow were filmed as well as Carmencita the famous Spanish dancer. In addition, Annabel Whitford was filmed doing the Butterfly Dance at the Black Widow and the film was copiedrighted.

In order to keep his invention in the making, Edison sought a patent on the Kinescope in 1894. He also established his company, the Edison Manufacturing Company in charge of the manufacture and sales of Kinescopes as well as film.

Kinetoscope sales declined when projected motion pictures started to take over peep show equipment in 1895. Also, competitors emerged that sold their own equipment for less than Edison's earnings. In part to make up for this, and to combat the decline in popularity of the Kinetograph the Kinetophone was released at the beginning of April in 1895. It fulfilled Edison's dream to connect the motion picture with the phonograph, and to make motion pictures talk. In order to operate the invention, a person gazed through the peephole viewer of a Kinetoscope, while listening to a music streamed through ear canals connected to the Phonograph within the cabinet. The device could not provide an exact synchronization system and was ultimately unable to establish an audience. The film that is now known in the present as Dickson

Experimental Sound Film is one of the very few examples available of this early attempt at sound (Bellis 2015).

The year 1898 was an active calendar to Edison as well as his business. Edison sent a cameraman to Cuba to film footage that were part of during the Spanish American War that year. While fighting on the home front, his son's second, Theodore Miller, was born. In addition, in this time Edison appointed Edwin S. Porter to assist on Edison's film gear. Edwin then became the most well-known director of the company.

Edison's film crew also ventured out into the fields and created documentary films. Some examples include Buffalo Bill's wild west shows. These films also featured Annie Oakley as well as the troupe of Native American dancers.

While Edison was busy and making films and his films, the two Frenchmen Louis and Auguste Lumiere created a compact camera called the Cinematographe. The machine was based on Edison's motion machine. The Lumieres projector projected films onto a screen, which widened the viewers of motion films. To fight those Lumieres, Edison obtained the rights to a projector similar to the one created by C. Francis Jenkins and Thomas Armat in 1896. In 1897, Edison presented his projectoscope, which was a modification of Jenkins's projector.

The early 1900s were the time when the Edison Manufacturing Company became a major manufacturer of equipment for motion pictures as well as films. In order to keep his venture into filmmaking alive, Edison created projection crews all over the globe to document interesting events and scenes. They were nothing more than travelogues, but they were fascinating and attracted people towards their first glimpse of the Black Maria.

Edwin S. Porter began to create films for Edison with more complex narratives. In December 1903 the hit movie, "The Great Train Robbery" was made and released. The subsequent films were made in Edison's studios, however in the same period, new film studios began to emerge and were competing against Edison's firm. In order to escape Edison's monopoly studios shifted the operations of their studios to Los Angeles.

Chapter 5: Edison's Little Well-Known Inventions

"Many of the failures in life have been experienced by those who didn't realize the extent to which they had come to success until they gave up." • Thomas Edison.

* Edison tried out an iron ore separator that was magnetic. The idea behind the invention was to use magnets to segregate iron ore from the low quality, unusable ores. He was involved in a massive mine project located in northwestern New Jersey designed to prove the iron ore separator can do. He spent enormous amounts of money for this research. Problems with engineering and the decrease in the cost of iron ore eventually led the invention to be discarded.

* Edison was aware that he needed to create a method to measure the electricity level that was utilized. He invented his Webermeter around 1881. The Webermeter was made up of four or two electrolytic cells that contained two electrodes that had zinc and an sulfate solution of zinc. The zinc was transferred across one electrode the next at a predetermined rate each time electricity was

applied. The meter reader took out electrolytic cells each time they needed to "read" the meters. The term "take" on the meters is still being used for electric power companies in the present.

* A talking doll? A fascinating invention, but these dolls were extremely bizarre and frightening. They sound like they were from another world. Little girls didn't want for them to interact with, or have these dolls. The production of these dolls was restricted to a small number And they were shut down within a year. Edison Toy Company shut down within an entire year.

Electric chairs were an invention by Thomas Edison. It was designed to demonstrate how risky Westinghouse along with Tesla's AC was. It was successful in its goal and Westinghouse suffered a lot of fame.

* Edison came up with an idea of conserving fruits, vegetables and other organic goods in glass. The glass vessel was filled with the food items to be preserved, and the air was pulled out of the glass using one of the air pumps. It was sealed using an additional piece of glass. Eureka! Food dehydrated.

Have you had the opportunity to play a recording player made of concrete? It was one of the tests made using foam concrete. The concrete was never advertised or sold.

* Then, Edison discovered that his ore separator could be used to grind rocks before 1904 when Edison patents his innovations in cement. He employed a rotary kiln to mix the ingredients and create a more smooth product. He sold cement via his business Edison Portland Cement. Edison revolutionized the industry of cement kilns by inventing it. The product of Edison Portland Cement was used for the construction of Yankee Stadium.

* He also applied his cement concepts to build solid, fireproof houses. Edison was planning to pour cement into huge wooden molds, which were with the dimensions and shapes of a home. He made ornaments, plumbing pipes and even bathtubs. The homes were advertised at $1200. The idea of concrete communities was not popular. The equipment and molds required for the construction of the homes required massive financial investments and there weren't many willing to embark on a construction project of this magnitude. There are still Edison concrete houses within New Jersey, but they are ugly.

In order to furnish the concrete homes Edison suggested that a home that was filled with concrete furniture crafted of air-impregnated spongy foam to keep them from getting too light was offered alongside the homes. To demonstrate his point, in 1911, Edison's company made the piano, bathtub, and cabinets. The furniture was slated to be revealed during New York City at the annual cement industry fair but the cabinets never showed on the stage and were never seen again (Barksdale 2014).

* Edison's work using concrete could be used for highway construction. Being the first pioneer to make application of concrete Edison contributed to the creation of America's first concrete roadways.

* In 1876 , another invention of Edison's was extremely useful. Edison invented an electric pen which later was later the source of inspiration for the first electrical tattoo machine.

* That fan with an electric motor which keeps you cool in the summer? It was the invention of Edison's.

* Did Edison create a device to let the lines open for communicating with spirits? Following WWI the spiritualist movement was

revived and many people believed that science would bring back loved loved ones. Edison believed in agnosticism however, he was not afraid to attempt to reach the spirits. But, Edison never introduced any machines of this kind and none was discovered after his death. Perhaps it was Edison's humorous side being broadcast to loud reporters.

Chapter 6: Businessman & Industrialist

"It's evident that we don't have a millionth of a percent about everything." Thomas Edison

Thomas Edison

In the following twenty years, Edison transformed into individualists, such as a marketer, industrialist, and business manager. His lab at West Orange was far too vast for one person to manage, and Edison hired a variety of scientists and engineers to develop his ideas. Edison was not so happy in his new job as businessman since he realized that some of his inventions produced and operated by mathematicians and scientists who had been trained at the university. Edison disliked the bustle and excitement of his lab and preferred working in a private and unstructured environment , with just a handful of reliable assistants. Edison realized, however, that he required an army of employees to help him turn his ideas into corporations.

Edison utilized a systematic approach to each invention. He was aware of the cost of his innovations had be economically competitive. Profit-seeking drove him to estimate the cost of each component in the process of developing his inventions. Once he was sure that there would be a profit then he accelerated the speed of inventing and manufacturing, then turned the focus on selling his inventions.

In a sign of his business-oriented tendencies, Edison developed a company that would supply electricity to power and illuminate cities across the globe. It was the Edison Illuminating Company, the first electric utility owned by investors that would later evolve into General Electric. General Electric corporation.

In the past, Edison was in contact together with Nikola Tesla who was an engineer with a degree in academics. Tesla was part of Edison's firm for about a year and then they publicly fought over how direct current power was used as opposed to the alternating current electricity. Then, Tesla entered into an agreement along with George Westinghouse who was an Edison competitor. A major dispute in the business world about electrical power was the result of Edison employed cruel and dangerous methods to convince people that the dangers of alternating current were real. The most well-known or famous experiments to demonstrate the dangers of alternating electricity included the electric severing of an elephant in the circus called Topsy in NYC's Coney Island. Edison also worked together with New York's State of New York to execute a prisoner who was on death row with the power of alternating current. This was done to further undermine Tesla along with Westinghouse. The results were disastrous because the technicians did not understand the AC voltage required for execution of the condemned prisoner. The initial jolt of electricity was not fatal to William Kemmler, but caused him to be severely injured. The procedure needed repeat. Reporters as well as George Westinghouse commented, "They could have done it better with an the axe" (McNichol (2006)). Edison however succeeded in creating anxiety around AC power, which caused Westinghouse to lose ground in his struggle for AC dominance.

In certain circles, Thomas Edison was reputed to be an industrialist with a cold blooded streak. He would go to any efforts to discredit his rivals such as he did during his Topsy the Elephant incident. He was never restful and continued to follow the same routines even when he was wealthy and powerful. He began every year with new ideas and attempted to transform his inventions into companies in order to dominate his competitors. Thomas Edison did create businesses and promptly apply for patents that varied from the making of cement to films batteries and the electronic lightbulb. He kept the patents close to him heart and was quick to bring lawsuits whenever other inventors stepped near his inventions.

Edison was surrounded by factories to manufacture his innovations and inventions. At one point, Edison employed more than 10,000 North Jersey workers in his labs. Edison always started his entrepreneurial initiatives with a state-of the art lab for invention, and maintained a tight grip on the development of ideas. It was believed that Edison could be tracked in and out as did his employees to efficiently assign his time to various projects. He was more active in the lab than the majority of his employees and only walked the half mile to his estate to visit his family.

In the 1890s, Edison constructed a magnetic iron ore process plant located within New Jersey. Edison was constantly looking for practical methods to meet the Pennsylvania steel mills' need of iron ore. Edison sold all his shares to General Electric to finance this project but was never capable of creating an extraction device that could remove iron from non-usable and low-grade ore. Then he gave up the idea but he was unable to recover the entire amount invested in this venture. Iron ore extraction from low-grade ores was an economic failure. Not one to allow anything fail, Edison salvaged the processes and came up with a better method of making cement.

He made his mark in his industrial days by creating an appropriate storage battery that could provide power to an electric vehicle. The battery Edison created to power the self-starter of Model T Ford. Model T Ford for his best friend Henry Ford was used extensively in the automobile industry for many years.

Edison developed a method to mass-produce similar wax-based cylinders. He also constructed the first indoor movie studio to be operated by Edison Company. Edison Company in New York.

In 1901, Edison took with him his cameras for the Pan-American Exposition in Buffalo, New York to prove to high-ranking officials including the president William McKinley, what his cameras could accomplish. The

Edison cameras weren't installed in the Temple of Music when President McKinley was killed by Leon Czolgosz. President McKinley was trying to shake hands with the anarchist at the time that two shots were fired. The cameras didn't capture the murder, however, a quick video of the people outside the Temple of Music indicates that the public was aware of what occurred. Film was used for documenting the funeral procession that was distributed to the public, helping to make cinema a news media.

Edison continued to work in motion picture production through the filming of "The Great Train Robbery, directed by Edwin S. Porter. In just a minutes, Edwin S. Porter was fired after he demanded the release of a "cut" in Edison's motion picture work. Soon, other directors replaced him without any hope of becoming wealthy. This was yet another illustration of Edison's intention to keep the profits local to his home.

In 1911 Edison's company was reorganized in Thomas A. Edison, Inc. The business became more diverse and structured , and Thomas Edison became less involved. Edison was the one who made decisions however, the objectives of the business were designed more to ensure market viability rather than to create.

In 1914 the year 1914, a fire was reported to have broken out at the West Orange laboratory complex and destroyed at least 13 structures. The cause was an explosion inside the Film Inspection Building. The damage was reported to be greater than seven million dollars. Only two million of them were insured. Edison as well as his spouse Mina were on the site and stayed until firefighters could control the flame. Fortunately , the Experimental Laboratory and Storage Battery structures were spared. Edison declared, "Although I am over 67 I'll begin all over from scratch tomorrow." (20th Century North 2008). More than 7000 workers reported for duty the next day . cleaning and reconstruction followed.

Chapter 7: Navy Career

"I am extremely proud to say that I've never invented weapons for killing." • Edison

Edison was not directly commissioned with the Navy However, there was an interest. He developed an electronic torpedo which could be moved on an electrical wire that could travel that could travel up to 200 miles from the host vessel. He recommended that the Navy use shells to fill them with calcium phosphate and calcium carbide in the Spanish American War. The shells were found to last for more than a couple of minutes after hitting. The aim was to light enemies ships. Edison believed that soaking soldiers in electric water streams could stop them from coming into contact.

He was involved in submarine project for the Navy which included the invention of submarine detectors as well as gun location methods.

Edison along with Edison's Assistant Miller Reese Hutchison worked many hours to encourage Edison the alkaline battery to the Navy to be used in submarines. In 1914, Hutchison had organized to invite Secretary of the Navy Josephus Daniels to meet Edison in the West Orange plant. Daniels was then invited by Edison as well as Hutchison to visit the battleship and submarine in the New York Navy Yard. After a few months, Daniels wired Edison, "Have just signed the authorization to authorize installing Edison batteries into the submarine L-eight." Edison replied, "I thought I was an optimistic, but your telegram will prompt the guys around here to put me on the equipment to stop me from flying." (Cronon 1986).

In 1915, following sinking the Lusitania by an German submarine Edison received a call from an New York Times correspondent to make a comment on the conflict. Edison said that [the government] must create "A huge research lab, in collaboration with naval and military and civilian control," the purpose would be to research and develop new weapons and defenses to ensure that, should war break out, "We count take advantage of the expertise gained through the research and produce large quantities of the most modern and effective tools of warfare." (Cronon 1986).).

Secretary of the Navy Josephus Daniels enlisted support from senior officials from the

Navy and the scientists. Edison accepted to serve as the president of the civilian experts of the Naval Consulting Board. The role of the experts was to advise the Navy on technology and science. He was not able to be the chairman as the chairman was not able to listen to the discussions. The assistant of the chairman recorded all the discussions of the Board by knee. People who were skeptical of Edison's capabilities said,

"He was, at seventy-eight and beyond, reading a few of the latest books appearing in the realm of pure science and asking intelligent questions about them as well. His ears were not there, but there was any crystallization of his thoughts that is typical of certain books before birth, in other cases, especially the so-called "men of action when we reach forty, and with the vast majority individuals ... before the age of seventy. The fact that Edison over all others retained his humility, modesty, and intellectual honesty and come to be known throughout his life, and even afterward and throughout his life, is, to me the most convincing proof we are able to judge his true extraordinary qualities. I'm referring of course to the ailment ...of public recognition and adulation" (Cronon 1986).

Between 1917 and 1918, Edison conducted special experiments in the service of the Government. Edison devised a method to determine the position of guns through sound ranging, and detecting submarines using sound waves from vessels moving, and an instrument to detect the release of submarine torpedoes. He also helped speed up the turning of ships , and also developed collision mats that could be used by submarines as well as ships. Edison was a researcher on ideas for stopping torpedoes using nets and also increased the power of Navy torpedoes.

He also created light sources for underwater exploration and sailing for convoys. Edison is believed to have camouflaged ships as well as water penetrating projectiles, (Beals, 1996).

The Board was in the initial planning stages of developing a new research facility. Congress had approved a grant of $15 million for the institution, however differences among Naval Consulting Board members and the near entrance of American to WWI delayed construction until 1920.

Edison was planning to build the lab with two distinct divisions: Radio and Sound. The two divisions were the pioneers in the fields of high-frequency radio as well as the underwater propagation of sound. Communication equipment such as direction finding devices Sonar sets, and the first equipment for radar were developed by this laboratory. The lab also conducted basic research and was involved in the early explorations of the Ionosphere. The laboratory was working towards its goal of becoming a global research facility. As World War II began, Physical Optics, Chemistry, Metallurgy, Mechanics and Electricity along with Internal Communications had been added to the lab (U.S. Naval Research, 2014).

Edison was a struggle to work within an organizational structure in which Edison wasn't the primary leader. Edison "butted" heads individuals on the Board and Navy management, but his years of working in the Navy were extremely productive. He was the source of inspiration to the Naval Consulting Board and taught the Navy to use the expertise of civilians to resolve Navy technical issues. Secretary Daniels admitted that , without Edison as a member of the Board and the industrial naval infrastructure of American could not be able to progress to the extent it did as it did in wartime.

Edison directly influenced Post-war Naval Research Laboratory, but Edison was strongly opposed to the lab being at Washington, D.C. As the result, he retracted his backing from the organization.

The Navy didn't immediately take advantage the full extent of Edison's inventions or even listen to his ideas although Edison was a major influence. It is said that the Navy wasn't the same following the encounter with Thomas Edison.

Chapter 8: Last Years

"It is stunning here ...."" Edison in his final resting place.

As the 20th century progressed, his health began to deteriorate and he started spending more of his time his home with his wife Mina. Edison didn't have an intimate connection with his kids. However, after he quit as the president at Thomas A. Edison, Inc. His son Charles was appointed as its president.

He sought the final 1 093 U.S. patents for an apparatus to hold objects in the process of electroplating. Then he moved to his winter residence located in Fort Myers, Florida where Mina and he Mina were able to work with the tycoon of automobiles, Henry Ford. Edison continued to pursue projects which ranged from electric trains to seeking an alternative to natural rubber.

Thomas Edison continued to experiment at home as the board was not in favor of him conducting experiments in his West Orange laboratory. He was determined to discover a substitute for rubber. These last attempts to find the rubber substitute were carried out on the advice of his close friends Henry Ford and Harvey Firestone. The rubber that was used in the manufacture of automobile tires was sourced from the rubber tree , which isn't a native of the United States. It was costly to import, and dangerous to heat and hard to mold. Edison experimented with a wide range of plants to discover the best substitute, and discovered a goldenrod weed which could produce enough rubber to be used. Edison was developing the rubber experiment when he died. death.

in 1928 Edison received his Congressional gold medal in recognition of his numerous contributions to the world. He was not done in his quest to discover and develop however. He started to create radio programs using long-playing discs. His first program disc that was successful was played by an radio station from New Jersey on April 4 1929. He founded a disc-phonograph manufacturing business, and later in 1927 his first Edison Needle Records was introduced. Edison was a little late to enter the world of electronic recording. Edison introduced a new line of needle or lateral thin shellac records that were compatible with any type of player However, the record came too late to be saved by Edison Records. The company shut down prior to 1929's stock market crash.

He made his last appearance an Golden Jubilee presented to him by his dear colleague Henry Ford. Ford rebuilt Edison's factory into a museum in Greenfield Village, Michigan to pay tribute to Edison's memory. Numerous notables were present at the event including the president Hoover, John D. Rockefeller, Jr., Marie Curie, Orville Wright and George Eastman. Edison however was unable to attend the entire event because of a health issue.

On October 18, 1931, at his house located in West Orange, New Jersey, Edison passed away. He was 84 and was suffering from diabetes for a long time. There is a legend that says his last breath was hidden in a test tube located at the Henry Ford Museum.

Numerous corporations and communities across the globe turned off their lights or shut off their power supply in honor of the passing of his father and to honor his life. Edison was the rags-to riches success story and a popular hero for Americans. He was an unrestrained individualist, a tyrant of employees and vicious to competitors. He was always in the spotlight but did not have a great social life. Edison was often rumored to ignore his family. But, when the time he passed away, he was among of the most famous and revered Americans around the globe. Edison is at the forefront of the technological revolution in America. Edison, along together with George Westinghouse and Nikola Tesla set the scene for the dawn of electricity.

In the time when America was in her most vulnerable in the early years of industrialization, Edison is credited with helping to build the American economy.

Chapter 9: Edison's Reborn

Mary Stilwell Edison's death Mary Stilwell Edison

The biographies written about Thomas Edison tend to dwell just a little bit on the topic of Mary Edison. Because of her self-assured nature and her habit of not paying attention to, or even not focusing on family and home life to focus solely on his creations, she does not appear in any biography of Edison's life in the early years. Newspaper reporters who flocked to Edison's side throughout the Menlo Park days did not often interact with her, or her children. the reporters who did get to meet her had very little things to write about her. This could be said about any wealthy housewife of the 19th century whom they wanted to be flattery to: she was thought of as attractive and beautiful as well as a committed mother and an ideal match for Edison. Journalists who wrote about her in any way tended to portray her as the comical stock character of the wife who is irritated and who's husband's inattention is a source of pleasure for her. This representation of

Edison is certainly based in the real world. Edison was known to be a bit distracted and was once seen at a Sunday dinner wearing an all-new, expensive wool suit that his wife bought for the occasion, only to walk out of the table and dive into weeks of unending work. He did not change his clothes, and the suit was evidently damaged. There is no evidence about Mary Edison survives, except in these sketchy caricatures. Edison seldom spoke about her, or did not mention her in his diary entries, except when he was attempting to attack her intelligence.

After they got married, Edison had taken Mary as an assistant at his lab but then abruptly changed his mind and sent her back to her home after she had been with him for just for a couple of days. "Dearly beloved wife, I can't think of anything something worth her time," he wrote in his diary in the year he was married. It's a thing that is not to be questioned, given that she was only 16 years old when they got married. wedding, and unlike Edison she was not playing with chemicals or conducting experiments on her own in train cars starting at the age of 12 to. Later, Edison was speaking to one of his colleagues who was about to get married. If Edison received a photo of the man's future bride, he casually applauded her appearance, and said, "Why is it...that so women aren't brainy? Brains of men are easily found however women are not so easy to find"

Edison and Mary were probably at a level of trust with each other to give birth to three children. However, it is difficult to not think that his constant absences from home was because of his insanity work ethic, or if Edison was not unhappy, whether unjustly or justly by the woman who he picked for himself. If he was disappointed it is unlikely to have blamed her for her decision. Given that he was 9 years older than Mary Edison, and had no contact with her when he decided to make a intention to marry her the likelihood is that he realized it was unfair for him to have done so.

Prior to dying, Mary Edison had been suffering from health issues. The exact nature of her condition is hard to establish. When Edison was traveling in Wyoming and California and during the third trimester of his wife's pregnancy, he was sent an email from one of his assistants telling the doctor that Mrs. Edison was very unwell:

"Mrs. E's health is not the greatest quality. She's extremely anxious and is constantly worried about you, and everything. I believe it's a nervous prostration. She was so scared last night by the possibility that the kids could get onto the track that she fell unconscious. This morning, I called the Dr. Ward who came at noon...She requires a change promptly, because the cars could keep her up in the night, which causes her to lose her strength."

"Nervous prolapse" was a largely non-specific diagnosis in the 19th century of medical practice. It was a broad range of symptoms resulting from a variety of unidentified problems. Women were believed to be particularly susceptible to the condition because the disorder was thought to be a sign of anxiety and emotional stress which was a sign of female vulnerability. The most commonly prescribed treatments for this condition were relaxation, changing of scene and "nerve tonics" that included medications like morphine and laudanum. There is no evidence about the treatment Mary Edison received for her illness, however her health was in a state of deterioration during the final few months of her lifespan. Edison did not seem to thought her to be in any risk. He was not willing to cut his journey to the west when he heard information about Mary's condition despite the fact that it was a fact that she was pregnant at the time. However, she was definitely in a state of decline. At the time of her death she had taken over the care of her Edison children for a time and likely wouldn't be required to perform in the event that Mary Edison were in good enough health to take care of the children herself.

After Edison completed the construction of the power grid on Pearl Street, he decided to step down from the invention aspect that was Edison Electric and commit himself to the routine business of around an entire year. When there was no longer a need in Edison to live in Manhattan every day, Mary Edison insisted that they relocate to their home in Menlo Park. In the spring of 1881, the Edison's daughter wrote to a close friend "I am so sick that I'm afraid... My head has splitting in two as well as my throat has become extremely painful." In 1881, the

Edison Family returned back to Menlo Park in the early summer. Two months later, on the 9th of August the 9th of August the 9th of August, 1881 Mary Stilwell Edison died. She was just 29 years old.

The circumstances surrounding Mary Edison's death remain a bit of a mystery medically speaking. The cause of death was described by the name of "congestion of the cerebral cortex" and a condition that as with "nervous prostration" isn't a clear definition in modern medical research. But new research into Thomas Edison's writings which is a program run through Rutgers University, suggests that Mary Edison's health issues were treated with morphine for a period of several years and it is believed that morphine played a part in her demise. At the turn of the century, morphine became accessible without a prescription and doctors could prescribe it to women who suffered from "nervous problems".

One newspaper, one of the very few who have ever published an interview with Mary Edison -- claimed, following her death the fact that she was an infrequent user of morphine and it was true that she passed away from a morphine overdose however there is no proof of whether the dose was taken in a deliberate or accidental manner or if it was administered by self or by a physician. The newspaper also stated that, just moments after the death of Mary Edison her husband tried to revive her with electrical shocks. Studies have revealed the fact that "congestion in the brain" as the cause of her death, was then regarded as to be a symptom that could be caused due to morphine overdose. It was also believed that electric shocks were often advised for "revive" patients.

Whatever the cause for Mary Edison's demise the fact that it occurred had a significant influence on Thomas Edison. It was more significant maybe than one could have imagined considering how far away their relationship was. Edison seemed overwhelmed with guilt like it suddenly appeared to him as if he had neglected his wife throughout her illness. Maybe he not believed there was anything really wrong with her but then he was proved to be completely wrong.

His daughter Marion known as Dot later remembered that she awoke in the following morning to discover her father "shaking from grief, weeping and crying so that it was difficult for him to inform me that Mother had passed away at evening." Marion Edison was 12 years old at the time, Her twin brother Tom was eight while her older twin brother Will aged five. The loss of his wife appeared to wake Edison to the reality that he was a father to children. He was never involved in their lives and sometimes he went so long without visiting them that they became obscure to him during the time between visits, however, that would change at least as far as Marion is concerned. Tom and Will who had been educated in boarding schools did not meet their father often over the years to come however, Marion was his main partner for a while. He resigned her out of Madam Mears's French Academy on Madison Avenue and took her home to stay with him. He bought her two parrots and a horse and brought her to the laboratory along with him with him as his assistant. (As an amusing joke, or perhaps to make her fit into an all-male atmosphere, Edison came up with an alternative name for her during her lab hours:

George.) He also assumed her education in an unusual way the lessons consisted in reading 10 pages of the encyclopedia each day. They also took carriage rides through the countryside in which Marion in charge of the horse, because Edison was not a great driver. Marion Edison was to remain with her father throughout the rest of her life until he married again when she was a teenager. she was to return to board schools.

The summer in 1885 Mina Miller and the Chautauqua Institute

Following the death of his wife, and after the electric light was placed in the commercial world and didn't require the supervision of his own, Edison began to feel that he was at ends. Edison decided to take a mandatory vacation from his work for the first time he'd done this since he was 12 years old. young.

In this summer, 1885 just a year following Mary Edison's death Edison went on vacation at the Chautauqua Institution in Chautauqua Lake, New York it was a retreat centre for leisure learning, similar to an actual summer camp that included educational lectures and activities for both children and adults. Edison was invited to join a session in the Institute and also to deliver lectures couple of years ago but had refused when he got to the final minute, and went to the west instead. However, now Edison was interested in the institute that he did not had before and that was that one of the institute's founders, Lewis Miller, had a daughter of 19 years old named Mina who Edison was planning to marry. Edison along with Mina met through the spouse of Edison's business associates He had been requesting the wife to present him to interested young ladies she believed he could be able to get along with. After that, she organized the opportunity for Edison to have a meeting with Mina Miller at a world's fair held in New Orleans (where Edison Electric had been awarded an agreement to light the principal structure) Edison was invited to make the trip to the Chautauqua Institute, which he delayed a couple of years earlier. Because Mina was also going and was invited, that he accepted and took his daughter Marion with him.

One of the things that happened during this camp was each guest keeping their own personal journals for 10 consecutive days. Edison was a regular keeper of diaries, however they were always more technical, with diagrams and sketches of his inventions and occasional remarks about the personal aspects of his life. However, his diary during his time at Chautauqua Institute Chautauqua Institute is of a distinct character. It is full of humorous observations and jokes about his work and the books he read and those he spent longest with such as the daughter of his and Mina Miller.

The diary exposes a facet that Thomas Edison otherwise obscured by the bare details in his autobiography. An individual so focused on his outward actions such as Edison didn't give readers an opportunity to look into his thoughts and feelings. Reading about the life of Edison from his youth to 38 is to be able to see the man who was always working on things, building things, and dreaming of ideas for new things to do and building. The diary of his summer of 1885 is to see the jovial, busy imaginative, tender mind which made all the doing or building. His style of writing had the richness that people might not expect from an individual who was so committed to science and technology; however, it is clear to discern, from the pages the fact that Edison was a long-time enthusiast of books and literature and literature, as well as his other achievements. The diary entry for the first day is found in the excerpt below:

"Menlo Park N.J., Sunday, July 12 1885

"Awakened in the early hours of 5:15 a.m. My eyes were
I was a bit irritated with the sunlight beams. I turned my back and attempted to take another dive into oblivion. Succeeded. I woke early at 7 a.m.
I was thinking Of Mina, Daisy, and Mamma G. Include all 3 of them into my mental kaleidoscope in order to find the new
combination a la Galton. Galton's style of combination. Mina as a starting point and tried to enhance her appearance by eliminating and incorporating certain aspects borrowed in Daisy as well as Mamma G. It was like a kind of Raphaelized beauty. I got caught up in the deep end, my mind went to sleep and I fell asleep once more.

"Awakened in the early hours of 8:15 a.m. Itching my head was intense with a lot of white dry and dandruff. It was a very uncomfortable experience.

What's this incomparable material? Maybe it's dust from the literary stale matter I've dumped into my nose lately. It's nomadicand is all over my clothes and I'm required to read about it within the Encyclopedia.

"Smoking excessively causes me to be anxious. It is time to sever my natural urge to adopt such behaviors. The constant snuffling of a heavy cigar inside my mouth distorted my upper lip. It is a bit of a Havana curvature.

"Arose around 9 am and got down the stairs, thinking it was too to be up for breakfast.
Twasn't. I couldn't eat much, my stomach nerves too and nicotinny. The tobacco plant's roots must be able to penetrate all the way into hell. Satan's main agent Dyspepsia should be in the control over this particular branch of the kingdom of plants.

"It is just dawned on my mind that your brain can digest specific portions of food like the ethereal component as well as stomach.

Maybe dandruff is an excreta of the mind, the amount of this material is directly proportional to the amount of reading one takes in. A book about German metaphysics, for instance, could ruin a dress.

"After breakfast, I began with reading the Hawthorne English Notebook. It's not a lot of fun. Maybe you're a barbarian of the literary arts. haven't yet been educated enough to appreciate exquisite writing. 90% of his work describes graveyards, old churches and coroners. He as well as Geo Selwyn should be appointed permanent coroners in London. Two wonderful things about the novel were. Hawthorne giving the tiny Rose Hawthorne a big live lobster, she told her that it was an ugly creature and that it would bite everyone, and she then asked "if the first lobster God created was bitten by him." Another time: "Ghostland is beyond the authority of truth."

"I believe that freckles on the skin are caused by a salts of Iron and sunlight helps bring them out by reducing them to a the high state to the low level of oxygenation. Perhaps, with a strong magnet for a period of time and followed by the right chemical treatments, these mudholes of beauty can be removed.

"Dot is currently taking care of the home of our dumb and deaf parrot. It has eaten plenty of food, but never made any sound from it. The bird is tame of a statue and the ability to produce dirt like an entire buffalo.

"This is definitely the most beautiful day of the season. It's neither too hot nor too cold. It's blooming on the peak of perfection, an Edenday. It's a perfect day to have an angel's picnic. You can eat lunch with the scent of fresh flowers and freshly mown hay, sip the fresh air while dancing to the sound of bees. Imagine the spirit of Plato riding on the back of a butterfly, cruising through Menlo Park with a lunch basket.

"Nature will smile at least once. Holzer has a puppy who just appeared on the patio. The dog's face was as depressing as a portrait of Dante However, the dog was waging its tail incessantly. This is clearly how dogs laugh. I'm wondering if dogs visit flowers to take a sniff. I don't think so. Flowers were never meant to be used by dogs, and maybe only occasionally for human beings, and clearly Darwin is correct. They are pretty enough to attract insects that serve as the transporters of their pollen, and pollen transportation via Bee lines.

"There is a nest of bumblebees in the vicinity of this veranda many times it has come close to me. A little bit of details (acquired by trial and error) I learned as an infant boy caused me to lose my interest in watching the movement of this flower burglar with arms.

"Had dinner at 3 p.m. The remains of a chicken and rice pudding. I am a fast eater.

"At 4 o'clock, Dot arrived on her horse "Colonel" and rode me around riding.
Beautiful roads. I saw a 10 acre parcel filled of] red raspberries that are cultivated. "A of a burying ground" in the sense of. I came across this hilarious pun on Dot. Dot declares she's going to write a noveland has she has already begun. She is one who is 16 but she is only twelve. We drove in the town of Metuchen. The town was named for the name of an Indian chief. They named Metuchen. Metuchen as the head of rolling land, the land

is uneven. Dot was astonished when I explained to her the idea of the church as a divine fire escape.

"Returned from the drive around 5 p.m. I began to reading short sketch sketches about the lives of Macauley, Sidney Smith, Dickens as well as Charlotte Bronte. Macauley was just 4 years old. young was an omnivorous reader. He used books in his children's conversations. When he was 5 years old the lady spilled warm coffee all over his legs. After a short time, they asked him if he felt healthier. He responded "Madam the pain has gone away." Her mother could have shaped his mind a few years prior to his body.

"Don't like Dickens I'm not sure the reason. I'll stock my literary collection with his work in the future. Charlotte Bronte was like DeQuincy What a lovely couple they must be. I have to be reading Jane Eyre.

"Played an hour or so at the piano. It's terribly off tune. The two keys are losing the voice.

"Dot has just read me the outline of her planned novel. The basis appears to involve a

wedding that is under pressure. I suggested to her that in the event of a marriage, to pour buckets of pain. This will make it more real. In terms of painting realism and other art forms, Steele
Macaye during a meal given at a dinner for H H Porter, Wm Winter and myself , we were told about a definition of modern realism, given by a Frenchman who's name I've forgotten: 'Realism, the dirty, long-haired artist standing in front of the statue of Shakespeare painting the old boot lined with dung. The bell rings for dinner. I head out."

The diary is hilarious and enjoyable It is also easy to comprehend why this is the case. it was designed to be read by the public. Every person at the institute who was part of the diary-keeping initiative was required to read other guest's contributions so Edison's remarks on Mina Miller were made in the expectation that she would be reading them as well as his comments about his daughter's wit and intelligence, as well as his views on marriage in general. Mina Miller must have been awed and ended up joining the Edisons and a different family in New Hampshire for the next phase during their vacation. On the way to the train ride across the mountain ranges, Edison set about teaching Miller how to communicate using dots and dashes of Morse code. This provided them with the chance to "speak" to one another in private, without any other passengers in the car being able comprehend what they were saying to each other.

It was a smart choice by Edison, as it allowed them to share a certain level of intimacy which most engaged couples could not enjoy but also allowed them to meet each other in a way they would have otherwise been unable to do because of Edison's hearing impairment. In their case, Morse was a substitute for the sign language. When Edison suggested to Miller just a few weeks later and asked him to tap the question on her shoulder in a room full of people. She responded with a tally of Y-E S before becoming agitated and left the room quickly which caused the confusion of her companion.

The moment that Edison was introduced to Mina Miller, his career was at a standstill. Reporters stopped coming to his door. They were able to tell that he had not fulfilled the extravagant promises he had made. The electric light was on however it had not substituted gas for gas overnight like he claimed that it could. Edison was seeking the next major project that would occupy his time and take his back to the exhausting work schedule he was relying on. He was unaware that Edison would never be able to duplicate the excitement and ingenuity that brought about the phonograph as well as the electric light, though both inventions would have an enormous impact on the subsequent decades of his existence.

However, his union with Mina Miller would reinvigorate his determination. Mina Miller was committed to the work of Edison, coordinating their lives in line with requirements of Edison's lab. Later, after Nikola Tesla had come to work for Edison He declared his opinion that Edison was so unorganized and unfocused in his decisions that he would have accomplished almost nothing accomplished without the knowledge that he was married to an exceptionally intelligent woman who managed his life with the greatest advantage. So far as it's possible to evaluate from a distance Edison's marriage to Mina was happier than the first one, perhaps due to the fact that he was a more loving spouse to Mina than he was to Mary.

A brand new home

After their wedding, Edison left it to Mina to decide if she would prefer to reside either in Manhattan and New Jersey. Mina picked New Jersey. Edison was able to acquire an extravagant, lavishly furnished house for less than less than a quarter of its worth because it was confiscated as restitution for an embezzlement trial. Edison and Mina called their new house Glenmont. After settling in with his family, Edison turned his attention to his own dream to create a private lab for himself, which matched that of his Menlo Park laboratory in the freedom and privacy it gave him, yet reflected his increasing wealth and status with its state of the latest design and equipment. Edison's dream for the lab was to have it equipped to "build everything, from a ladies' watch to an entire locomotive." The expression was often repeated even though Edison never constructed watches for women or locomotives, the phrase served as a symbol of the extent of his creativity throughout his life.

The initial idea was to create a new research corporation around the lab , and then draw investors to finance it. However, Edison had earned a reputation for being a difficult person due to his interactions in the company's board Edison Electric. The board wanted Edison to remain at the lab and stay clear of the business aspect of things. Edison desired complete autonomy over his business and even took cost of earning more money than would have been possible by putting the matter to competent management. He was beginning to realize that he could be more content by leaving the business side of his company to real business people however, in the meantime investors were becoming uneasy about working with Edison. Edison ended in paying for the lab through his own pocket and became the sole owner.

Edison was the inventor of a variety of new inventions that He wanted to get started on in his laboratory, which was situated within West Orange, New Jersey and included long-dead ideas like the hearing aid as well as innovative ideas such as the automated cotton-picker he had imagined during his honeymoon. He certainly didn't wish to go back to the phonograph or the electric lights as they have long since ceased to captivate the attention of his. However, circumstances forced him to get back to the phonograph after 1879 for the first time.

The main issue in Edison's phonograph - the problem that was preventing the sale of it for commercial use was the fact the fact that the cylinders or record, on which the phonographs played, was made of tinfoil. This was not able to be removed of the phonograph, without harming the cylinder. Also, the phonographs were able to only play the cylinder on which they were bought, which was quite different from Edison's dream of selling phonographs containing thousands of cylinders of music.

The only models of phonographs Edison had ever released were models designed for use in traveling exhibitions, in which only one single record could ever be played because the audience would vary each night. Since Edison had abandoned the phonograph nevertheless, Alexander Graham Bell, the inventor of the telephone was able to create the graphophone, a device that was functionally identical to the phonograph, with the exception that it replaced wax cylinders with foil ones. The wax cylinders were able to be removed and replaced with new ones which made the graphophone the first machine that was that could play music at any time.

The graphophone's appearance came as a sour shock to Edison who, even though Edison had lost interest about his creation, couldn't be bothered by the thought of another person improving it and presenting it to the public before he'd made it available to the public. He was too competitive and possessive to allow this. It was the American Graphophone Company made overtures towards him, and acknowledged Edison as the inventor of the phonograph and offering substantial shares of the company if they could utilize his name. Edison turned down the offer and told the investors of his company that he was working on an improved design of the wax-cylinder that was in the process of development. When he set out to show his model to investors there was a problem. According to Alfred Tate, one of Edison's laboratory assistants:

"Edison was confused. There was no method to explain the result. He tried repeatedly to get the instrument to talk, but time and again, it would only yell at him. The time allotted to our guests was extremely limited. They were given one hour to the demonstration. They had plenty of time to use the instrument. It was clear that Mr. Dolan and Mr. Cochrane needed to get on an express train to get home at Philadelphia and their scheduled time for departure was set when Edison was still in an attempt to replicate his voice. The gentlemen kindly said they would promise that they would return to the laboratory after Edison discovered and fixed the flaw on the machine. They then left. They never returned."

The issue was caused by an error that an Edison's laboratory assistant done to a component of the machine that didn't work as it should. The damage was already caused, but Edison could not attract more investors to create an aggressive collection of phonographs which could rival graphophones. He was forced to enter the business of a smaller firm that was producing graphophones in addition to phonographs. And even at that point, he did not have the same level of success.

The earliest the phonographs that Edison's North American Phonograph Company were in poor condition; they were frequently broken and the cylinders of wax were brittle and susceptible to cracking. In addition, Edison was still determined to promote the device as an office tool that could be used for dictation purposes and reducing the work of Typists or an amanuensis. However, the phonograph was not the best choice to be used for this, since most people did not want to be completely still and speak into the tube of the phonograph as they made notes. Edison himself did not make use of the phonograph to do this. The device could reproduce the sound of a human voice in a way that was understandable as well as could reproduce the music's sound much better, however Edison did not yet have a firm grasp on the idea of producing Phonographs solely as entertainment devices.

Samuel Insull

A period of intense sadness followed Edison's failure in the race to launch the phonograph. The phonograph, as well as the electric lights he had created were far from the success it was promised they'd be at the end of the 1870's. Edison grimly joked with his personal assistant and secretary, Samuel Insull, that it was his intention to return working to earn a income in the field of telegraph operators like he had been when he was younger. Instead, he appointed Insull his manager of business, which was the reason it fell under Insull that Edison's General Electric Company--still known to the world as General Electric -- would eventually become a reality.

Samuel Insull had come to work for Edison from his home in England at the age of twenty-one years old. He was extremely gifted, and equally notable for his self-confidence exaggerated in business as his status as the only one on Edison's staff who was familiar with him enough to be able to avoid making fun of his. Insull was the only one to move to America in the United States after Edison promised him the position of his personal secretary, a post of exceptional responsibility for someone of his age. Edison did not just grant him the post, but he accepted Insull to heart and leaned on him like the only employee, with the exception of perhaps Charles Batchelor, who was more mature than him and had been with him for longer.

The trust and trust among Insull with Edison were mutual. When Insull was given the charge of the brand new Edison Machine Works plant in Schenectady, New York, He was required to accept a cut in pay, in addition to having to throw extravagant events for Edison's customers from his own pockets. He decided to not speak something to Edison about it, believing that it was a mistake that would be fixed at some point.

When Insull completed the first report of his annual year to Edison the company, he had happy news to report that was a vastly increased sales rate in every product line and timely payment of bills and an efficient use of professional letterheads. Under his direction, Edison Machine Works began to make gains of 100 thousand dollars per year, which was a huge improvement. Edison must have realized the debt he owed to Insull's management because the company promptly gave an offer of $75,000 worth of shares in the company and when he realized that Insull was paying the cost of entertaining customers as well as the maintenance of an equines stable that they could hunt, Edison gave him a significant pay increase and assured him that in the future, there would be separate fund that would be devoted to entertainment costs.

Edison was still at his laboratory located in West Orange, New Jersey He rarely had the time to travel to the Schenectady plant, which over the space of just six years, the workforce been increased from 2100 people to 8000. In the words of Insull himself said, "We never made a cent until we built the factory situated 180 miles of Mr. Edison." Relations between the West Orange and Schenectady sides of the enterprise did not always go as planned. Edison through his laboratory worked alongside the machines works plant like every business that Edison was solely responsible for the labs were not efficiently managed from a commercial viewpoint. The bills were never paid and Edison often made the machine works plant pay twice for lab services that was charged to anyone else and that Insull was extremely annoyed by. However, Insull was able to work with Edison for several years regardless.

Chapter 10: Edison At War

War of the Currents: The War of the Currents

Edison's electric lighting company did not have the same success than its competitors this fact puzzled and irritated Edison. The biggest issue in selling Edison's lights to consumers was the same as before the fact that he couldn't put in power grids to supply power to vast metropolitan areas, without incurring massive cost. The competition had found solutions to this issue but through a former Edison employee named from Nikola Tesla.

Tesla was an Croatian inventor who had been brought to Edison's notice by way of Charles Batchelor, who oversaw Edison's telephone exchange in Paris during that time Tesla was employed there. Tesla was able to an instant spark of inspiration, come up with the idea to construct a safe and long-lasting alternating current induction motor. This was something that Edison thought was not feasible. Tesla was just a teenager and when he spoke to Batchelor the engineer, he was keen to travel towards America. United States and explain his amazing concept to the famous and well-known Edison personally. Batchelor was extremely impressed by Tesla and offered him a letter introduction, which would guarantee his employment in one Edison's American offices. However, he advised Tesla that Edison would not like to hear anything more about alternating current. He'd met only failure when he tried his own experiments with AC electricity, so was determined to do no more to have to do with it.

Tesla realized this was the case when he arrived in his home in the U.S. and met Edison personally, however he remained in a friendly manner with Edison for a time, despite. Tesla quit Edison's business due to an issue with money. Edison had made a joke , promising Tesla an enormous amount of money if he could resolve a glitch with one of his electric motors. Tesla spent a whole one-year period of his life to resolving the issue, and Edison tried to give him a increase and promotion, however Tesla was unable to keep up with Edison. He decided to leave Edison's company and start a the field for himself. This is how he came across George Westinghouse. Contrary to Edison, Westinghouse was deeply fascinated by Tesla's alternating current induction motor.

The reason the power of alternating currents easy to distribute over a larger area, unlike Edison's direct current which was limited to smaller grids such as those in the Pearl Street station area--was the transformer. It was able to take the brilliantly bright light arcs that were used on the streets to provide outdoor lighting and let them be turned up and down in brightness, which allowed them to be dimmed down enough to allow them to be used indoors. It made transfer of power much more efficient and, consequently, much more cost-effective over the long term.

Edison at the time of the 1880's was being told by his top executives to think that Edison Electric needed to develop its own version of an alternating current system in order to stay competitive in the market. However, Edison was a stubborn man in this regard which no one could argue with. First of all whether unjustly or not, Edison felt personally betrayed by Tesla. It was a resentment to him that his former colleague had adopted the technology that Edison was against and gave it to Edison's competitor. Additionally, he held personal views about the business of electric lighting generally. The idea of the tech evolving differently to the one the one that had been developed by him caused Edison desire to bring it back on the original path. The third reason was that he was convinced that electricity with alternating current is extremely risky, in a way direct current was not.

There was a certain amount of scientific evidence to this theory which we explored in the previous chapter, every electrical current causes muscles spasms in the body. However, the alternating current can cause fibrillation that is extremely hazardous. It is also true that there were many deaths due to accidents during the initial days of electrical technology, because people didn't yet comprehend how to take appropriate precautions when they encountered live wires. However, Edison did not take the time to understand Tesla's idea before dismissing the idea. Tesla's motivation to develop the induction motor with alternating current resulted from his observation of an example of Direct Current

Gramme dynamo during the engineering college he attended in Austria. Tesla was aware that it was sparking in a dangerous way, and spent the rest of his life attempting to come to a solution this issue. However, Edison could not see it as the answer to any problem.

The War of the Currents is the title given to the war of propaganda that was fought between Edison's firm and the electric companies that alternating currents, Westinghouse in particular. The gas companies used the tactics of sabotage and false information to attempt to keep people off from electricity. Edison was able to respond by highlighting all the dangers related to gas, the explosions, fires and poisonous fumes, the unpleasant smells and the acrid colors. Death, as he implied it was a real danger for those who had not switched from electric to gas.

By the end of the 1880's the gas companies had no longer been considered to be a significant threat to Edison's business, not because they were defeated, but due to the fact that the majority of customers use a variety of electric and gas fixtures. Electric companies that used the alternating current method, nevertheless, posed a significant threat, and Edison's propaganda apparatus responded to the situation. According to Tesla Biographer Margaret Cheney puts it, "accidents caused by AC should be avoided if they can't be identified be manufactured and the general public notified of the dangers. Not only were lives at stake during the War of the Currents but as well the personal glory of an individualistic genius."

Many stories started appearing in newspapers about the shocking death of people injured by accidental contact made with electrical wires. The following example was published by The New York Times on January 21st 1887. The article is titled, "Struck Dead In A Second":

"Shortly at 5 o'clock on the afternoon, as Vesey-street was at its busiest state, flames were observed emanating through the basement at No. 49, a three-story brick structure that is used as a shop for William Wilson & Co. that operates as the Centennial American Tea Company, and are known to have an extensive inventory of coffee and tea.

It was just on Wednesday that neighbors noticed a huge amount of tea arriving at No. 49, there was some curiosity noticed in the matter about the fire. Prior to the alarm being responded to the fire, it had gained a considerable amount of ground and the heat of the street's narrowness was extremely hot. The flow between the Eighth and Sixth avenues street vehicles through the main thoroughfare was stopped and Vesey-street was handed over to firefighters.

"There was a huge, cumbersome canopy at the to the front of the shop. it was discovered that wiring that were connected to the United States Illuminating Company running across the awning would be a major obstacle to the firemen's efforts. Director Fred Simmons, of the United States Illuminating Company, was observed at the intersection of the road. He was summoned and given the job of cutting wires. It was an attractive young and athletic, and attractive man who was around 34 years old, however, the job seemed to be easy and safe that bystanders initially were not interested in the work. He set his ladder on the ground, then walked quickly up the slopes until his shoulders were an even level with the wires. He carried a pair of the pliers in his hand. As he slipped one arm across the the ladder lean forward, and then using his other hand, cut the wire in two.

"As he was doing this, an illuminated arc shot out for a moment before it faded away. The man's body was shaking in its elevated position, then expanded to a full size. A scream was heard from the spectators who were able to discern instantly that the Superintendent was receiving all the current of electrical fluid that was flowing across the wires. In a short time, nothing was accomplished. Then , one of the firefighters recovered his mind, leapt onto the shed, pulled the wires the Superintendent was still holding, and the body fell to the ground, with the impact of his head when it hit the pavement was clearly heard by the spectators.

'"...The clothes that Mr. Simmons was unburned, his facial features were not at all disfigured, and the death could be the result of heart disease. So free is the look of the corpse. any signs of painful illness. [...] the ambulance doctor stood up and examined the body.

'"Why you're dead?"he declared. "This isn't an instance for this particular Chambers-street Hospital.'

'"What was the cause of his death?', asked a passer-by.

'"This isn't the case,"" said the surgeon in a snarky tone and then he ran away."

In this tale as well as others that were published throughout the 1880's, one can feel a tangible fear and wonder at electrical energy. It appeared to in killing without leaving an visible sign that it had been absorbed into the body. Edison was not happy with this. The things that people couldn't be able to see, they could not be afraid of as readily as if they had obvious burns and wounds. It was also essential for Edison that people be afraid of alternating current. So crucial that he was to create the term the alternating current, as well as its principal advocate, George Westinghouse, synonymous with death.

The electric chair

A few days after the story about the death the Superintendent Simmons from the United States Illuminating Company appeared in the New York Times, an even more alarming article was published, titled "To Eliminate Hanging The commission will make a report on the benefits for electricity." The article reads:

"The document from the Capital Punishment Commission will be presented to the Legislature on Tuesday the following day. It is expected that Mr. Elbridge T. Gerry who has been in Europe for a while sent a message to his coadjutors should he not arrive in time to discuss on their behalf, they must submit a preliminary report, and request a further period of time. The Dr. Southwick, of this city, a co-member of the committee went to Albany today to meet with Matthew Hale, the third member. Matthew Hale, the third member. Before he left, the Dr. Southwick was asked what the results of the work of the commission will be. He responded:

'"The amount of opinions expressed in the responses that the commission received in the circular it sent an email to prominent attorneys, judges as well as other members of the State, seeking their views on the issue is in opposition to hanging and in favor of electric. The report will be in support of the introduction of some electric devices to execute. This is the final goal toward which I've been working for the past six years. If this report from our committee doesn't result in the passing of a bill that abolishes hanging, I'll be tempted to conclude that I've worked in unsuccessfully. I've noticed that the bill passed by our Legislature this year was copied in Paris and an identical bill was presented by the Frenchman within the legislative body of France. Germany has also taken on the issue and I've just discovered the news that New Jersey attention has been attracted to our efforts to raise the issue. I would like to see New Jersey would be the Empire State would take the initiative to move in this direction towards broad human rights. One argument which could be argued in favor hanging is its deterrent power, however I believe that a peaceful death would be able to have the same effect on society, even if carried out in the privacy of. It is possible for a prisoner to be kept to an State prison and then be released from his life in a quiet and peaceful manner without the hysteria and shiver that comes with the hanging."

The notion of the electrical chair being invented as it was believed to be a painless method of execution could be a surprise for people of today, who are likely to be familiar with the horror stories directly from execution rooms. Electrocution, a mix consisting of "electrify" in addition to "execute"--as it is carried out by prisons who execute capital punishments for prisoners is not in any way like the Superintendent Simmons who died in a single, clean death. Prisoners who die in electrocution are basically burned to death. Their bodies burn as their blood boils and their skin is ripped off from their bones. There isn't always an instant death. It is reasonable to claim the time that Edison was approached for assistance in devising the method of killing prisoners using electricity and he did to make it as painless as possible.

He also did all that he could to ensure that the electric chair function for his own purposes of propaganda. When the commission addressed Edison to ask for his suggestions, he suggested they make use of an alternating current device manufactured by George Westinghouse. His colleagues even went as in suggesting using the term Westinghouse could be used as the term used to describe the operation of the machine. In other words, instead of saying that prisoners were electrocuted, we could declare that they were "westinghoused". What better way to cement that in the minds of people that electricity generated by alternating current could be dangerous than connect it with executions?

The name was not picked up on, however the procedure was scheduled to be tested in the trial of William Kemmler of Buffalo, New York. Kemmler's lawyer argued obviously that the state's decision to apply an untested and untested procedure for execution is in violation of his client's rights. But who was able to say for sure that the execution would go as simple like Edison claimed? Judges who heard the appeal, deposed several experts which included Edison who was adamant about the power required for the execution of an adult. The judge was not aware that Edison was not entirely certain of the facts. Over the course of a few weeks, he was testing the voltages required for the execution of animals with electric current. According to the historian Margaret Cheney puts it, "Edison was offering schoolboys 25 cents per head for cats and dogs that he later electrocuted in intentionally crude experiments using the use of alternating current. In addition, he distributed scare-leaflets that contained the words 'WARNING!' in red letters on highest point." The death of these animals fulfilled two purposes: research into electrocution as well as propaganda against Westinghouse. Regardless of the fact the fact that direct current was just as effective in killing people, but the actual fact that an alternating current killed these people was evidence that it was unsafe to allow within the homes of.

Westinghouse was sick of watching himself and his company slammed in the media. Edison's comments were beginning to cross the line of personal insults to professional ones that Westinghouse was unable to tolerate, especially because he had made personal appeals to Edison but was rejected. As per Edison biography Randall Stross, Westinghouse published an article in the North American Review answering some of Edison's most damaging assertions concerning alternating current:

"Westinghouse also dug up some old conversations with Edison and discovered this quote"I don't really care as much about a fortune, but I am concerned about outsmarting the other peers.' Westinghouse suggested to readers that it was this, and not the supposed advantages of Edison's own system which prompted Edison to overstate the dangers inherent in other systems and minimize the risks for his system."

General Electric

According to the views of many historians according to the opinions of most historians, it appears that American consumers were not particularly concerned by wars of Currents. They didn't have any strong opinions about how direct current was more secure than alternate current. They prefer alternating current five times one over Edison's direct-current because it was less expensive and was more easily accessible. After it became obvious that Edison lost his battle in the War of the Currents, Edison started to pull out of his own business. Edison General Electric united Edison Electric with Edison Machine Works and all the other smaller businesses that were created solely to supply Edison Electric with the parts that it required for its manufacturing. Edison sold 90 percent of his stake in the company which netted him around three million dollars in cash.

Samuel Insull, Edison's faithful right-hand man, was appointed vice president. In 1892, he was in charge of the merger of Edison General Electric along with another electrical company called Thomson-Houston He resigned as president of the Thomson-Houston business shortly after. General Electric, the new business General Electric, was the first to not be named after Edison. It's not clear if Edison was the one who chose this or not. A source states that Edison was adamant about not allowing the name of his father to be linked with the company. However, his children were likely to assert later on that later in life, it was a source for intense sadness for Edison because his name was taken from the business he been working to establish.

Mining

Ogden, New Jersey, was the location that was the site of an iron mining facility that Edison bought in 1889, just a few years prior to leaving his electric business. Edison was the first to have the idea of mining during his travels across Wyoming and California but he never had the time or the funds to explore his ideas fully after he had sold his stake of General Electric, however, Edison had funds and the time to pursue whatever he desired. The thing he sought most of all seemed to be the chance to explore something that was completely novel to him. His most enjoyable part of the invention process was always the initial year, where ideas exploded quickly and huge leaps were made in a matter of hours. The new inventions held an influence on his mind that the routine process of creating an affordable machine to sell commercially never did.

Edison was a miner for five years. the mine. He resided at the Ogden campsite for six of seven days during the week or more. The camp conditions were extremely harsh. He as well as his team mates were housed in a plain home made of clapboard (nicknamed"the "White house") without any amenities and no shelter from elements. The enthusiasm he had for the mine was likely to be high enough for him to remain there in these conditions, since his wedding with Mina Miller was still relatively fresh, and their bond was constant. He wrote an extensive number of correspondences to Mina during his time in the mine. They expose a side of his character that is not found in any of the personal documents of his, except that it is mentioned in his diary kept during his time at the Chautauqua Institute, at the beginning of Mina's relationship. In these letters, he refers to Mina as "Billie"--his tendency to give nicknames to female family members remained constant.

"August 9 of 1895. Darling Darling Billy Edison and 2 angels on top. Today's temperatures were more humid than the seventh section of hades that was reserved to Methodist ministers. It was humid to the point that it became so intense that certain fish in the Hopewell pounds swam to the sky. The mill's dust was terrifying, and it caused people to leave The White House. The cows left us but we had a fantastic morning... It's Saturday tomorrow, and I'm feeling disoriented that I'm not back home to visit my sweet clean Billy. What do I do without a bath? the seeds of smartweed have begun to grow out of my jacket's seams. Mallory is planning to pay an order for a bouquet of seeds for flowering to plant in my clothes. Imagine it this way, Billy sweetheart, your beloved changed into a garden of flowers. I'll feel lonely in my home on the next day. I would do everything to have you and the kids here. They have written 2 letters before this. I wish you an excellent rest and be content. With every dust fleck today , as a counterpoint for the kiss. I am your love unchangeable apart from strangers."

"August 11 1895: Darling, Billie E. It's been hot over the past two days. You went out just at the right time. It's raining hard now. I'm doing well, and if I weren't too busy, I would feel lonely. Are you able to find a boy in this world that is similar to Charles? I'm sure you'll travel quite a distance before you can find one... Tonight, I was feeling like a sandstorm without the love of yours. With a kiss that resembles the swish from a 12 inch projectile from a cannon, I am forever your love."

"August 12 1895: Sweetest, Darling Lovely Cutest Additional Billie Edison [...] you are the most adorable thing in the world and why would you ever need to suffer the blues. There is no reason to be blue other than perhaps regretting having a love like me. ..."

"August 15th 1895: Dear Billy dear, don't be down about yourself. There's not a woman in 20000 that is truly as smart as you. Their apparent intelligence is insignificant due to their abilities to talk. Their judgement isn't worth anything Your insecurity is the issue. Being blue about such issues is untrue. Check out the newspaper's and be daring Billy and stop reading novels. Enjoy as much as you can. Everything is wonderful, and you're with a person who is loving you ever more every day. You don't have to worry about his reliability and constancy. Therefore, kiss your darlings 21 times for me and keep the normal amount of your own."

The mine for iron ore wasn't a huge success but Edison was awed by the effort involved in it tremendously. When he read in the paper that the shares of General Electric were trading at an all-time high He asked a close friend to calculate the value of his shares would be worth if he been able to hold them. The person he asked came up with 4 million bucks. It was reported that Edison was very sombre for a while, but then his face became bright and he said, "Well, we had great fun with the money." Edison had invested it primarily in mining, so the pleasure he felt from the work must have been awe-inspiring. Even though the value of the iron ore mine was much lower than initial estimates suggested, the project was not completely lost The equipment Edison developed to process the iron ore in his unique method proved to be useful in the new field of making cement.

Columbia Exposition

With the invention of the phonograph Edison was the first to introduce the world to the possibilities of recording sound. At the time of the 1892 Chicago World's Fair, he was making his first attempts at motion pictures that were that were recorded and synchronized with the sound. This World's Fair, also called the Columbia Exposition, was held to celebrate the 500th anniversary Christopher Columbus discovering the New World. It was intended to display all that was great about America, particularly in the areas of technological advancement and invention. The entire structure that was devoted to the marvels of electric lighting and the known as White City (a temporary village consisting of whitewashed façades) shined brightly in the arc lights that were outside which hung down the canals and avenues.

Newspaper reporters were asked by newspaper reporters Edison to confirm if he was adding anything new to the show and he reverted back to his previous habits of speaking to the press, making reference to and explaining the invention which did not necessarily existed: "My intention is to be able to create such a wonderful combination of electricity and photography that a person can sit in his home and watch on the curtain the silhouettes of opera players on a distant stage and listen to the voices of singing artists." This was in reference the kinetoscope and the device that could do "for the Eye the same thing that a phonograph does to the Ear"--only it couldn't perform the extent Edison claimed. In addition, everything it was able to do was thanks to his chief photographer and assistant, William Kennedy Laurie Dickson.

The kinetoscope operated with the same fundamental concept as the phonograph in that it operated on an hand crank, and then turned a wax cylinder which produced sound. There was another cylinder that contained hundreds of tiny, hand-mounted images that were able to pass through the lens of something resembling an ocular for a microscope, and when the device worked correctly it was able to play music and images that would move in perfect the same time. It was an innovative device however Edison would like it to function similar to television sets that were color however, something it was unable to achieve. Dickson was given the task of developing a new design of the Kinetoscope, just in just in time to be used at the Columbia Exposition, but all did he manage to accomplish was to work himself into exhausted that he was forced to cease working for a while. The kinetoscope wouldn't be completed until a full year after the event was completed.

Stross analyzes Edison's attitudes toward the kinetoscope in this manner:

"It was evident to all however Edison that the Kinetoscope when it was finally ready for release, would prove to be an immense source of enjoyment all kinds: the absurd as well as the spectacular and the ridiculous. Even before the kinetoscope had been launched it was reported that an Albany newspaper published a report that it would be to record boxing matches, which would allow thousands of people to watch the match in just one week after the event. Edison was able to instruct the general public in a pious voice on the machine's potential to record performances in the Metropolitan Opera House in New York."

The Kinetoscope and movie projectors.

It was often said about Edison that he was a man who knew little about enjoyment as people knew what fun was. He was a huge believer in work being enjoyable. He was genuinely excited about creating, which is why he could say the reason for his long working hours of 18 hours. For what was to come in the entertainment industry it is true that Dickson as well as Edison's other laboratory assistants could have known a little more about how a typical person was having fun. After constructing one of the very first proto movie studios inside Edison's laboratory the haphazard structure was named "Black Maria" following the police vehicles of the time They shot a number of short films using the Kinetoscope. The first film featured an athlete showing the muscles he had (his normal fee for appearance was two hundred fifty dollars, however he decided to waive the fee when he had the chance to meet the famous inventor Thomas Edison). The second film featured the Spanish dancer who was displaying a shocking amount of legs, and the third movie depicted an infamous fighting cockfight.

Kinetoscopes were quickly installed in shops and patrons bought tickets to enjoy watching the tubes of rubber and watching the show as if they were in the theatre. The machines eventually were fitted with slot machines which became the first coin-operated entertainment device, and the grand-mothers to arcade games. Coin operated kinetoscopes generated incredible profits and quickly attracted the attention from two men named of Grey and Otway Latham, and their colleague Enoch Rector, who contacted Edison's representatives in April of 1894, offering that they would utilize the kinetoscope's technology to film and display boxing matches.

The technology of the day restricted the tales that the Kinetoscope's creators could tell using their equipment. It could only play a single 20 second reel at single time and, even when multiple Kinetoscopes were set together to create an extended narrative the machine was unable to create a complete story. However, Enoch Rector was able to increase the capacity of the kinetoscope in size from twenty seconds reels up to 60 second ones that could include three minutes of boxing in the event that the film was speeded up. (One of the fighters who was asked to participate in an opponent prior to Edison's cameras noted that he could have beaten his opponent more quickly but didn't want to be moving too fast for Edison's equipment to keep up with.) When the match was shown at the time of its 1895 debut, every one of the six rounds were shown on an entirely different machine. Visitors paid ten cents each to view only one round, and sixty to view all. The queues to view the kinetoscopes were so long that police had be brought to ensure that the audience was in control.

Edison was at this point mostly occupied with his mining of iron ore in Ogden and, while he was satisfied with the work that his lab did on the kinetoscope at the time, he was less concerned about the next major idea that everyone was thinking of to improve the kinetoscope. The demand for kinetoscopes had been so large that Dickson and Latham, Lathams, Rector, and the others all realized they could earn a lot greater profits if they could could find ways to get images from the box and onto the screen. Kinetoscopes were boxes that measured four or five feet high that had eyes that let you look at the image. Customers had to stay in their seats, sitting in awkward positions, to be able to watch the spectacle. If people could sit for a show, more people are likely to purchase tickets. If the image could be projected onto an omnipresent screen tickets could be sold to as many as the space could accommodate. Edison did not want to think about creating a projector, as he did not think it was feasible without impacting the quality of images.

Other inventors were also developing the projection system in Edison's shadow. The issue with the image quality was due to the amount of time that it required for light to travel over an image. The Latham brothers solved the issue by making the images bigger and developed the projector they named the pantoptikon. The name was changed to the Eidoloscope within a short time, and it began to show public displays of "life large" prizefights, which attracted huge crowds. (Boxing was prohibited within New Jersey and several other states; Edison had narrowly escaped being investigated for his role in filming a prizefight in his own property. It is no doubt that the illicit excitement was a major factor in the ticket sale.) The most significant improvement to the quality of images came from two inventors, Francis Armat and C. Francis Jenkins who recognized that light had to be allowed longer time to travel through the images. Their projection device, known as the Phantascope was able to pause each frame so that the light could be saturated it, which resulted in more clear projection onto the screen.

It was the Armat as well as Jenkin's "screen machine" that eventually was sold under Edison's name. The investors of the company had pleaded with for him to create a projector that was his own invention, but his response was the same enthusiasm that he has always displayed when investors asked for things that were not in line with his interests in the present. He didhowever decide that it was best to sell it under the Edison name, which was the first time in his career--never previously had he let an improvement made by someone else on the invention of one of him to be branded with his name, even though it was the American

Graphophone Company had once offered him the same offer. The device was eventually available to the general public in the form of the Viscoscope.

In the year 1897 Edison was the main character of a twenty-second film reel, titled"Mr. Edison At Work In His Chemical Laboratory, produced in the Black Maria studio. He was seen for the entire film moving around the fake laboratory like he was busy working on an experiment. It's likely to suppose that the modern audience who were completely unfamiliar with the concept of films and filmmaking, were not aware that the lab was staged or that Edison was not working on any kind of experiment when he was filming. If the average viewer was more knowledgeable about the process of making films the audience had no reason to doubt the film they were watching. In the reel Edison wears a huge white lab coat and Edison never glances at the camera. He is too busy shifting faux chemicals across tables, setting them on top of burners, and then analyzing the results with a serious expression. The image that he created of himself, the distracted genius engaged in a plethora of scientific issues that anyone would be unable to comprehend is in complete harmony with the image he always shown to newspapers. It was an impressive piece of advertising and it helps in explaining the lasting reputation of his name.

Chapter 11: The Edison Name

"The public's fascination with Thomas Edison's inventions grew and fell when announcements of upcoming wonders would attract attention, but then delays in the release of these marvels would be disappointing. In time, however Edison's fame gained an unbreakable sheath, and surpassed the attention given to inventions in themselves. The person who was most admired by the public is Thomas Edison, the person who the world was most attached throughout his time in existence. Edison was aware of this, and was tirelessly worked to safeguard the most clear image of his personality his name. "Thomas Edison' was a recognizable name. Edison is an exemplary invention too."
Randall Stross, The Wizard of Menlo Park

Fake Edisons

While Edison's most popular invention was his fame however, it was as likely like his other inventions to being snatched up by people who were opportunists. In the 1890s, Edison was selling his names of his invention to producers of a myriad of innovative inventions he favored however had no involvement in the creation of, for the reason that machines or anything else with the Edison name on it proved to be a huge success. However, there were many other Edisons who were named Edison all over the world and a few of them realized that they could offer names and rights names to makers who believed in the assumption that consumers will not be too concerned about whether the Edison endorser of their product was actually an inventor of fame. Thomas Edison brought suit against the fake Edisons. The suit stopped for a time the fraud, but shortly Edison found himself in competition with the most surprising source: his son.

Thomas Alva Edison, Jr. was popularly known to his family as Junior for his entire family aspired to create inventions just like his father. Being the second child from Edison's original marriage he'd not seen much of his father as young and even less after Mary Edison's passing and his divorce. Edison went to boarding school when he was an teen, but he left without graduating , as Edison was determined to start his own journey being an inventor. Edison offered him work in the West Orange laboratory and at the Ogden mining site However, Tom Edison was never able to attain the ease of interaction with his father he had hoped for. It is unclear if Edison's hearing impairment hindered them from communicating effectively or just Tom's shyness. Tom appears to have enjoyed an affectionate relationship with his stepmother, Mina but she wrote him an email in May 1897 in the Ogden camp in which he conveyed the frustration that he was feeling in his famed father's business:

"My dear Mother--I had hope--'though I'm not sure the amount'--that you would be able to please me with a single word or two. I'm sure that's true. just been away from 'Glenmont for a couple of days, but these days! What a long time they go by--I will sit and wait with that sense of complete contentment and the feeling of anticipation that sometimes occurs, though it isn't often. am waiting for those words that are far closer to me than the author has ever or will ever.

"But why should I bother asking you to speak for yourself, knowing that I'm not worthy of one single word? Because you are the one I ask mother, you, and no anyone else--and every mother would never cut off her son from her advice or the wisdom she has. Father has graciously given me a task to write down that really delights me, and I am working hard at it. I hope that it will be a hit with him. I try my best to achieve, even though I never have the ability to please him as I'm sure it's not me.

"But I will not give up if I could only speak with him in the way I would like to, maybe things could be different. I have lots of thoughts of my own that, at times might be able to say at any time I'd like to inquire about or share with him however they do not disappear from my mind and then are forgotten--maybe to be put in the right place, or perhaps not. That is what I'd like to see him decide. Well my mother, I don't have the desire to bother you for a long time, but I thought it would be nice to take the initiative if I felt it to be necessary. It's not that important to let you be aware that I am still around, and am thinking of the people that I love, just as I always do."

The following time, the reporter for The New York Herald wrote a article about Tom named "Edison Jr., Wizard", that included a variety of extravagant claims about his work as an inventor, which fit well in the Edison story, but with no foundation in reality for example, the assertion that he created an electric lightbulb that was superior to the one his father had. The journalist accepted the claims of Tom without conducting any research to prove them The consequence was that Tom was an overnight sensation on his own. The reporter was asked to serve as the host for the Electrical Exhibition at Madison Square Garden in 1898. Afterwards, people were allowed to believe that he was the one responsible for the display of electric lights they saw. In reality, his primary responsibility was to create an appealing layout for the light's arrangement. The promoters of the event just wanted to to claim that Edison was a fan of the exhibit. Tom was compensated for his contribution, however another lucrative chance came his way.

Tom Edison began to be approached by investors looking to create companies that had his name as a part of it. The first came The Edison Jr. Steel and Iron Process Company and later Thomas A. Edison, Jr. Steel and Iron Process Company. Thomas A. Edison, Jr. and the William Holzer Steel and Iron Process Company. Tom was awarded three-quarters of the shares from the previous company and was promoted to vice-president of the second. However, both went under quickly. Tom was unhappy that he was unable to not obtain loans from banks since when bankers saw that he was a banker, they thought that he couldn't possibly have money in the bank. Wasn't his father famously wealthy already? Tom was adamant to tell the father of his that, if he'd shown any expertise as a businessman or had left his business operations to people who were more adept at the task then he would be millionaire "ten times over". While it was true that the Edison family was relatively wealthy, Tom was anything but an unspoiled trust fund child.

After Tom's business ventures were scuttled, resulting in his exile from the city by debt collectors, he got approached by the Edison

Chemical Company, which manufactured ink for commercial printers. Edison Chemical was named after an individual named C.M. Edison who did not have anything to offer the company, other than being the name for Edison, the famous inventor. They were removed from business for a short time through a suit brought by Thomas Edison, but that didn't hinder them from negotiating an offer with Thomas Jr.--it was an act to trick the public into believing that their products were endorsed by Edison however what if they were able to claim to people that their products were approved by his son? Tom was delighted to accept the deal, not just because he needed cash, but also because the time had come for him to develop inventions of his own to make available to the public, such as one called the "Magno Electronic Vitalizer" that claimed to treat a variety of awe-inspiring illnesses.

Clearly, Edison took the rebranded Thomas A. Edison, Jr. Chemical Company to the court and won his case a second time. He made a variety of insulting remarks towards his son during the process, and claimed that he did not have any skills as an inventor. Tom however was quickly shattered by the pressure the anger of his dad and accepted a lump sum from the father in exchange for never selling his title to his name in the future. In fact, he altered his name from Thomas Willard changed to Thomas Willard, as if thinking he had renounced the rights of being an Edison. He was so angry Edison Sr. over the entire incident that he wrote to the younger brother of Tom, telling him to inform Tom that he did not want to ever see him ever again. The brutality of this paternal judgement appeared to have a profound influence on Edison's oldest son, who struggled throughout his life with alcoholism and depression.

The automotive and music industries

There were two initiatives that were the most prominently in Edison's mind during the latter half of 1890 and into the early 1900's. One was connected to music and the second to batteries for cars. This was the one he was most interested in. Edison was working in the design of a battery which could power an electric vehicle and he hoped that it would cost less for people of average income than keeping horses. In the same time advancements in the technology of the phonograph were contributing to the growth of the recording music industry. Victor and Columbia are still the giants in the music industry today began after it was discovered that disc-shaped records held twice as much of music than the original cylinder for phonographs (four minutes instead of two minutes.) The first time that musicians could record their work and earning popularity and earning money from sales of their music. The single therefore predated the album, and it would take a few years before the recording technology could store more than one track on a record at one time.

It is possible to be excused to believe that a man with the hearing impairments of Edison wouldn't have a great opinions about music. He could hear music however, he could only hear it with difficulty. Edison told a newspaper that while listening on the phonograph he needed to put his ear directly against the wooden cabinet and in the event that this didn't work it was necessary to sink his teeth in the wooden cabinet in order to capture the sound waves that passed through the mouth. Edison had a variety of opinions on music, however, Edison had opinions about everythingand was not afraid to present himself as an expert on any area that caught his attention, or was connected to the inventions he had invented. Edison had clear thoughts regarding what kind of music is worth recording and selling and what kind of music was beneath him and consequently, beneath the company he ran. His music standards were individualistic; he resented jazz, however he disliked certain composers that performed in the concert halls such as Sergei Rachmaninoff. He asserted in a somewhat bizarre manner that true music enthusiasts were drawn to "soft" music and that only people who enjoyed "soft music" were likely to become frequent, loyal buyers of phonograph recordings. Those who favored jazz or other "fads" were likely to be one-time customers in a way, not capable of supporting the record industry overall.

In reality, the rules for success in the business of music were exactly the same in the 1890's and in the present: certain songs become popular, which result in massive sales for a short time. The profits from some of the most successful songs were enough to pay for thousands of other songs that didn't achieve a meteoric popularity. Edison was not a person with a deep knowledge of the factors that influenced the market for his invention. He always seemed to be holding on to his personal notion of how people should behave, just like when he claimed that the phonograph worked better in the office for taking dictation rather than an instrument for entertainment. He was particularly upset by the celebrity culture that was created around music performers who signed agreements with companies that made phonographs; perhaps due to his frustration over his fame. Edison felt that attention to the performers was distracting from the attention that needed to be given to the music. He tried to stop record companies from adding the names of musicians to the music recordings of their music. However, the era of the famous music producer was upon him and he couldn't resist it for very long.

Henry Ford

One of the repercussions of Edison's popularity was that he didn't have many close relationships with his friends. He was naturally skeptical of the motives of any who contacted him, particularly when they were businesspeople or were involved in the field of the field of technology or inventing. He believed that he'd suffered in the past by those who relied on their connections to him to create private deals that were detrimental to his interests. This could be the reason why he was unhappy about the scandal involving Tom Jr. and the Edison Chemical Company. To establish an actual friendship with someone, the person needed to be an equal, someone with whom he shared things in common, and who was sufficiently successful to be successful in their own right to require anything from him in terms of his fame or technology. In the year Edison reached his 60s and a similar person came to him in the person in the form of Henry Ford, inventor of the Model T car.

Ford was just 16 years Edison's senior, and he admiration for Edison when the time he was a teenager working for the Edison electric lighting companies. They once met, prior to when Ford became famous at a conference in Michigan. Ford was giving a talk about his idea to develop the engine's internal combustion. According Ford, according to Ford, Edison was so engaged in his lecture, he exchanged seats with a person closer to the front of the room to ensure that he could better hear and, after asking many questions then declared that Ford had discovered the solution to making automobiles affordable and easy to access. Edison was known for his refusal to encourage other inventors, or acknowledge that they had an idea that was good and this praise from his hero, affected Ford greatly. Ford believed that Edison's support was instrumental in providing him with the motivation to continue to work till the engine's internal combustion was developed.

They didn't meet another time until 1912, when that the Model T was in production and Ford was almost as well-known than Edison was. Ford was in negotiations with Edison's secretaries over months until Edison accepted to allow him to come to his lab. From then on they became fast acquaintances. Ford being aware of Edison's enthusiasm for electric vehicles wanted him to create an electrical system to be fitted into Ford automobiles.

Edison stated that he'd be extremely attracted by such a venture however, the cost was too high to finance it on his own. He could visit Wall Street in search of investors but was worried that his credibility with the business world was and he would be unable to repay it. Ford was himself adamantly opposed to Wall Street, and his deep affection and admiration for Edison was so great that he decided to join Edison's financial team on the battery-development program.

Being busy and famous, Edison and Ford did not spend much time socializing on the streets, but in the latter years during his lifetime, Edison probably saw more of Ford than any other person who wasn't a part of his family or working in the workshop. The families of both took vacations together and Ford lavished Edisons with presents. He sent them numerous automobiles from his factories as well as dealerships, not just towards Edison himself but also to his children. At this stage in Edison's life, he did not offer much to Ford apart from acceptance and friendship. The advantage in this friendship was largely transferred between Ford to Edison and not the reverse.

This became evident in time, when the battery Edison was developing to support the Ford company did not work. Edison did, as was his usual, diverted his attention from the battery work to concentrate on the production of the phonograph but even after he had set the batteries as his primary focus the model was not able to work. The most striking result that came out of Ford and Edison's business partnership was the fact that their friendship was able to endure despite the failing. Ford isn't shocked by the fact that Edison failed to keep his word on his promises or to have been apathetic that Edison could not pay back his loans on time. Following a catastrophic fire in Edison's lab, Ford offered him one million dollars to repair the lab; and in 1925, as Edison's health was declining, Ford forgave the outstanding amount of the business loan he had taken out.

Ford famously said Ford famously said that Edison was the most brilliant inventor on earth and also the most unprofessional businessman, because Edison did not understand or care about business. The comment was that was made of Edison numerous times throughout his lifetime, however from Ford the criticism was warm. To Ford, Edison deserved every honor solely because of his ability at inventing. The failures of his business didn't diminish his genius or in any manner. Edison and Ford experienced more, but than less, of each other when their business partnership was shattered.

Edison's last years in the company

Edison was ill-healthy in his declining years. He was diabetic, and was suffering from stomach problems which were not diagnosed. They could be caused by the early, hazardous experiments using x-rays (his partner in the x-ray experiments was diagnosed with skin cancer, having to have his arm amputated before succumbing to an early death.) But, even when he offered his children Charles and Theodore the positions of a limited amount of accountability within Thomas A. Edison, Inc. He did not want to leave and hand the company over to the two sons. Through the course of his existence, he believed that he was at the threshold of his next breakthrough in the world.

In 1915 when the United States on the verge of joining World War I, Edison was invited to join the newly-formed Naval Consulting Board, in which scientists from civilians suggested technological advancements that could give to the U.S. navy advantages in war. Edison was fully in the board's hands. Onboard the U.S.S. Sachem which was a private boat bought by the navy to conduct Edison's research and research, he analyzed the possibility of various equipment that could aid in camouflaging the position of gun and ship vessels and even detect the location of torpedoes. The U.S. Park Service, "Edison was expected to spend 18 months in the field and would think of an array of forty-eight distinct ideas, such as the hydrogen-detecting alarm, which would help to prevent the risk of explosions undersea vaseline and zinc antirust coatings to protect submarine gunners, as well as an antiroll system for vessels' that would ensure precision even in rough waters." His military consultancy career was not a happy one but the navy did not use a single one of his ideas. It was not an insult against Edison but out of the more than 11,000 ideas proposed from the Naval

Consulting Board, only around a hundred were thought of for implementation by the navy and only one actually saw use during the war.

In the course of Edison's wartime service during the war, his son Charles who was the director of the company, established an entirely new department for personnel with the aim of which was to enhance the lives of employees working at Edison's manufacturing facilities. Charles cut the working hours from 12 hours to 10 hours, and established an on-site hospital with a doctor and an assistant in the event of accidents at work. He also instituted a worker's compensation scheme, despite suggestions that it could cause the company to go under. After the conflict, Edison took full control of the company once more and wiped out any modifications Charles was making. When the company began be a victim of the Great Depression, he singlehandedly dismissed seven thousand of the company's 10000 employees. According to the legend the legend, he would wander through the hallways of a firm and challenge employees with a series of stern questions regarding the nature of their work and if they didn't answer in a manner that was satisfactory to him, he would dismiss employees on the spot.

In 1929 in 1929, at the time that Edison was just eighty-two the bulb that was incandescent marked its 50th anniversary. A huge celebration, the Light's Golden Jubilee, was arranged by Henry Ford, and included notable guests like the president Herbert Hoover, scientist Marie Curie who was the creator of the plane Orville Wright, and billionaires John D. Rockefeller Jr. and J.P. Morgan. The ceremony that was live broadcast on radio during which Edison with Ford and Hoover recreated the moment when the first incandescent lamp exploded to life. "Mr. Edison is carrying two cables in his hands," said the broadcaster who was narrating the story. "Now Edison is reaching for the old lamp, and Edison is connecting the two wires. The lamp is lit! The Light's Golden Jubilee has come to the climax!"

Edison lived two more years following this ceremony. In the beginning of 1931, Edison was granted his last patent, which brought the total number of patents he filed in the time over his lifetime to one thousand ninety-three. The following year, he experienced kidney failure, but was able to recover for the next couple of months. In the end, on the 18th of October, 1931 after spending a number of days with a coma that was intermittent, Thomas Edison died at the home of Glenmont his home, which it was purchased at a bargain forty years ago as a wedding gift to Mina, his spouse. Mina. He was in the company of his family and wife.

The day Edison passed away the obituary of his death was published within The New York Times under the title, "Human Qualities of the Inventor and various aspects of His Exciting Life":

"Thomas Alva Edison made the world a better place which to live , and also brought high-end luxury to the lives of the average worker. There is no one on the list of those who contributed to humanity more to make life easier and comfortable. By inventing electronic light, he provided the world a brand new brilliance as the cylinder that was the first phonograph recorded sound, he made the music of the past within the reach of everyone When he invented the motion picture , it was a gift to the world of a theater that was an entirely new way of enjoying. His inventions provided jobs in addition to illumination and entertainment to millions.

"His creative genius sat in the midst of the world that at nightfall was in darkness and only pierced by the weak beams of kerosene lampsor by gas light bulbs or in the bigger urban areas, the uncertainty of the old-fashioned arc light. For Edison and his vision of the incandescent lamp lingering in his thoughts it appeared that the people were still living during that Dark Ages. However, his ferreting fingers were sifting through the dark until they brought out the glow which told him that the lamp's incandescent light was a hit, and that the goal of light for all was achieved.

"Thus he allowed other people to continue his work in this field of study however it was because of those early breakthroughs that America is the leader in screen effects, and it is the pennies arcade which featured a shooter's gallery as well as knockout battle films, has given way to the cinema's cathedrals. Additionally, thanks to Edison the possibility exists for the inhabitants of Kamchatka to be able to watch in silence for hours, row after row, and observe how the champion diving team from Rural Centre, Ill. held an annual water festival and raised funds to pay the church mortgage. In turn, it is possible for the children from Rural Centre to observe how the well-behaved native of Bengal is doing in the event that a hungry leopard tries to eat him. Edison was more than just a light for the lamp in Menlo Park."

Chapter 12: Edison In Manhattan

Trains and ships

Following his New Year's celebration in Menlo Park, Edison was appointed to equip the luxurious steamship Columbia with electric lighting. This was the first time Edison's lighting was used outside of the city of Menlo Park or his own laboratory, however Edison was keen to get involved in the challenge. In the years following, when Edison Electric was a global company with branches throughout the world Edison's lights were in great demand for ships, because the public had an unquenchable desire for the beautiful sight of an electric ship floating down the ocean in darkness. And The Columbia was the vessel that sparked the fascination. Lights on ships with electric bulbs could become dangerous once they became more prevalent, since wiring could be susceptible to sparking and fires on ships can be dangerous, but the illumination of the Columbia was a huge successful one for Edison. On the first voyage of the ship through Cape Horn, however, passengers were prohibited from turning lighting in the cabins off or on They had to notify an steward each morning and at night to open a lockbox close to the door, and then operate the switch.

In another instance, showing his unending ability to let new passions to divert his attention from projects in the process of development, Edison took time away from constructing commercial versions of his electric light to be used with electrical trains around 1880. Edison believed that an electrified train track with an asphalt layer to keep it grounded and shield it against water damage, could be less dangerous and more speedy than other train. Furthermore, it could be completely automated and controlled via a telegraph signal. Because the majority of train accidents were because of human error, such as when Edison was only 16 and was unable to stop an approaching freight vehicle on an accident course with a train--removing the requirement for human engineers to be a great idea. The electrification of rails would keep the train's wheels from running off the track, the second most common reason for accidents. The electrification of rails would allow trains to run at quicker speeds than those driven by steam or coal. Edison was so ambitious as to construct an electric train track measuring half one mile long in the fields surrounding Menlo Park. His visitors, he as well as his lab employees tried the train out by riding the train themselves often for up to forty miles at a time until it did leap the tracks and slam certain passengers in the fields of grass around the tracks.

Competition

The delay that Edison took in bringing out commercial versions of the phonograph was unlikely to cause him any harm since it was in fact his creation and nobody could contest his patents. He had the advantage of taking as long as he desired to bring the phonograph into commercial production, since there was no chance of anyone threatening him with the phonograph, amateurs working off the drawings and specifications of Scientific American regardless. Electric light production was different story. Scientists as well as engineers, inventors and scientists were working on the electric light for many years before Edison was even born. there were many different approaches to solving the issue of creating an efficient, long-lasting electric light. Even though Edison's method was the most successful, the secrets to his success was discovered by the Herald article on Carbonized Paper filaments.

The same reporter who wrote in the Herald report, Edwin Fox, wrote Edison an open letter in October 1880, in which he urged Edison to put lighting made of electricity to commercial use within a short time. Through the windows of Fox's office, he had a view into the labs of the newly formed United States Electric Light Company in which they were making bulbs based upon Edison's idea, while in a narrow way, skirting patent issues by bending the filament of cardboard into an M shape in honor of its creator Hiram Maxim. (Edison's bulbs were bent to the form of horse shoes.)

In fact new electric light companies were popping up across New York, and most of these were beat Edison to the to the. Edison was planning to electrify the entirety of lower Manhattan however, at the same time, he was researching bamboo as a possible new filament material. He sent his staff on trips to every country that grew bamboo around the globe to collect samples so that Edison could decide which was the most effective variety (he ultimately settled to Japanese bamboo.) There were certain advantages to using bamboo filament however it wasn't able to change the fact that even though Edison was still working on his electric lighting, United States Electric had already established an office in Manhattan with a reading area with one hundred fifty electric bulbs. Edison believed he could effortlessly slide into a deal with the city in order to electrify Manhattan however, his business was losing ground each day to his rivals. Brush Electric Light Company had already agreed to provide electricity for Broadway for free to the city, purely to benefit from becoming the very first electrical light business in the city.

Edison was always aware that his most frequent clients were private business owners. There was no incandescent light bulbs which worked well for outdoor use thus street lighting was always an arc light that was not the type of technology Edison was able to build his name around. Then he realized that if he didn't seek the public works contract for lighting it would give up an enormous advantage to his rivals. Additionally, Edison was already having difficulties trying to negotiate with the city about the right to construct the wiring grid required before he could provide commercial customers with electric lighting. To try to reach a fair deal for the wiring grid, Edison put on an extravagant light show in Menlo Park once again, this time as a private event held for all the New York City aldermen who were the sole decision-makers on the conditions of the contract between Edison and the city. The light show were followed by a lavishly served dinner, complete with champagne. The aldermen returned back to the city in a great mood however, when negotiations were resumed in the weeks following, Edison balked at the terms they provided. They demanded that he pay ten cents per inch of cable laid that, if you consider the amount of wire required to power the city, could have cost Edison more than 1,000 dollars for each mile. They eventually reached a compromise of $5 per foot.

Leave Menlo Park

Edison's laboratory in Menlo Park was his ideal workplace for a number of reasons. He had worked for years to build it into an idyllic place for eccentric genius inventors. It was close enough New York to be convenient yet far enough that he didn't get as surrounded by curious pedestrians than he might would have been. He valued his independence when it came to work above any other thing which is why in Menlo Park he had it. But, at the beginning of 1881 the lawyer he was working with recommended urgently shifting the entire operation and the family he was raising and his family to Manhattan. The company could buy an office building for its offices. The building could electrify completely, which could be the most effective advertising opportunity to the New Yorkers about what Edison's light could accomplish. Edison himself would need to be present to ensure that the effect was fully realized, which would require living in New York. Edison knew that this mean that he would have to interact with more people who were idle than he was afflicted with in Menlo Park, and that the man would need to commit an even greater amount of time in demonstrations and answering frequent questions. However, he was convinced that this was beneficial to the business, so it was decided that the Fifth Avenue offices of Edison's Electric Light Company became his new home base.

Edison was then required to buy the land in Manhattan to house the dynamo that he was required to construct for the electric light which he planned to sell. However, even in the year 1881 Manhattan is a prohibitively pricey place and even when he decided to buy the most decrepit buildings he could come across in the worst part of town he had ever heard of Edison had to pay 150 times more than that he had planned to spend. To top it all off it was necessary to establish an entirely new corporation to build the dynamo since the shareholders of Edison Electric were not keen to expand their stakes in this direction. Edison established his own company, the Edison Machine Works Company as the result.

The dangers of electric power

Edison was told from the town of New York that his workmen were to be under the supervision by five safety inspectors who were responsible for ensuring that electric wiring was laid safe. Edison was responsible for the cost of paying the inspectors. He didn't mind the expense , but he was worried that task would become stalled because of the safety concerns of the inspectors. However, as it was, he really was not worried about anything. The safety inspectors never visited Edison for a brief minutes on paydays and left Edison's workers at their own.

However, the safety of electricity was becoming an issue of concern for everyone. Today, everybody is taught since they're young to be wary of electricity as well as not touching wires that are exposed, and not to stick objects made of metal or fingers into electrical outlets. There was no general knowledge of how electricity operated in 1881 and even though electrical engineers were conscious of the need to be cautious but uninitiated visitors to factories and labs for first time weren't, and electricians did not often think to alert them. As wiring for electric appliances became more prevalent the number of deaths resulting from electrical accidents naturally increased.

Edison Electric has not yet had no fatal accidents among its staff or visitors however, some of their competitors did, and Edison determined to assure that the public that he operated exclusively with direct current electricity unlike the other companies involved in the deaths, which worked using alternate current. This was the very first ploy of what would eventually come to be called "the War of the Currents" which was a long and bizarrely violent public relations war that pitted Edison against the most well-known historical adversary, Nikola Tesla.

Edison Biographer Randall Stross provides this useful explanation of the distinction in direct and alternate current

"Direct flow of current is in the exact direction. the current that alternating is flowing in one direction, and then reverses and flows through the opposite direction, constantly changing. Both kinds of currents can electrocute a human which causes a continual muscle contraction. Alternating current is a particularly risk of injury due to its fast and abrupt flow--whether in this direction or than the other--is more likely to confuse the neural subsystem which acts as the heart's metronome that guides it. If the signals become confused the fibrillation will follow: fast and ineffective contractions of heart muscles that fail to pump blood the way it is supposed to. The tendency of the alternating current to trigger fibrillation provides direct current with an advantage in terms of security."

The most common sight in major American cities in the beginnings was horses acting oddly around electric wiring - rearing, bolting and then accelerating at a speed of a gallop, much which was a source of annoyance for their drivers. There were times when people and women who walked down the street could feel either their toes, or entire bodies, being irritated. This was caused by electric "leaks" which were places in which the insulation of the wiring was defective which was causing electricity to flow to the ground, which carried the charge with a slight twang. Edison was first experiencing issues similar to this within the Pearl Street station, but Edison was not open with the public regarding them particularly after journalists poured into his office, demanding explanations and explanations. Edison's staff went as they claimed that, not only was there no accidents occurred, but that an accident of this kind could not even be imagined. It was actually contrary, as Edison realized all too well.

Gas and electricity vs. electricity

Prior to that War of the Currents, there was a conflict between gas and electrical companies. In the previous chapter that gas was a powerful monopoly during the 1880's. Once it started to see Edison as well as other electrical light companies as direct threats to the business of its clients, they didn't be afraid to resort to untruthful methods in order to convince people that if they switched their homes from electricity to gas, they were in danger of dying. It was the same company that sent a saboteur in Edison's lab with instructions to short-circuit Edison's incandescent lamps for a demonstration. However, this was actually an simple gesture when compared to the lengths they willing to go to maintain their monopoly.

Edison and various electrical corporations responded insisting on bringing attention to each gas explosion that took place as well as every suspected death which could be attributed the poisoning of gas. They published a plethora of books on the subject, which included pamphlets that examined the amount of oxygen in rooms with one person reading a book by gaslight to be comparable to the oxygen levels in a room with no gas, with 23 people. Gas was undisputedly the most harmful fuel. has a distinct smell that affected the furnishings and walls and released fumes in contrast, electricity was tasteless and odorless and didn't emit heat. However, comparing the disadvantages of gas with the advantages of electricity wasn't enough. Electrical companies wanted to make it appear with the general public's perception that using gas within their homes meant a courtship the death penalty. The gas firms' marketing claimed similar things about electrical companies. The assertions that both sides made were flims at best or outright lie at the worst. However, the purpose of the gas companies was to keep control of the biggest utility in the country. The magnitude of the money involved led to people being extremely ruthless.

In contrast, the electrical firms' objective was to promote a new product as well as, in a more general sense, a brand new concept for the American population. Because the characteristics of electricity were obscure to the general public, allowed gas companies to make hints at all kinds of fatal effects of using it however, at the simultaneously, they enabled electrical companies to offer a variety of dazzling or bizarre assertions about the benefits of electricity. The readers of newspapers grew used to hearing about electricity being acknowledged for a variety of improvement in physical health as well as mental health and appearance.

The article below, "A New Use of Electricity" was published within the New York Times on January 12th 1882. It's a great example of tongue-in cheek reactions of the times to the variety of consumer products that were saturating the market after being electrified in one manner or another, and to the outrageous assertions made by advertisers on behalf of these goods:

"The use of electricity is expanding every day, mainly the applications made of it by clever advertising. The hair-brush that is electric, and is reputed to allow hair to grow in the hair of brass monkeys when it is used regularly it has been in the hands of the public for a long time but was until recently widely regarded as the most efficient and simple method of applying electricity on the skin. The hair-brush is, however being surpassed in the opinion of the general public with the corset that is electric. It is a amazing and innovative invention. The wood-cut that shows the method of using an corset electrically represents the article as a type of close-fitting jacket worn by a woman of a certain age who's neck and sleeves of which are slightly too alarming? The wood-cut is essential because, even though the advertiser tell the general public that the electric corset looks identical to the normal corset however, the words of his advertisement convey no notion to any upright and decent man. If worn for a long time this corset with electricity will allow the wearer to become large and experience the best of health - that is, if we believe the claims of the advertiser. Since it claims to be totally safe, and not causing any impact to the human body the corset should become at a minimum as well-liked as the hairbrush that is powered by electricity is. In spite of the advantages of the corset that is electric it is believed that a breakthrough is being made with regards to the electric hairbrush, which, once it is made public is likely to make it to be the most sought-after item for women to keep in her home.

"Like numerous other remarkable discoveries, this one at hand was found through accident. It's been a part of the culture that the slipper worn by the mother is the device through that nursery discipline is enforced. It was certainly true years ago, when slippers were worn by all and a lot of the most cherished memories of childhood for men of the current generation are tied to the slippers worn by their mothers. However, the traditional slipper that could be put on the foot and put on in the areas that would perform the best at any moment however, has been largely been discarded. The boot with a button is now the new standard and, it is not just impossible for a devoted mother of a large family to take off and then re-button her boots a dozen times in the course of a day however, the boot itself is too bulky and coarse an instrument to use to teach oral lessons. This is why the hairbrush has become the most popular method of teaching children to be taught in a proper manner. It's always within reach, it comes with a comfortable handle and its back is broad and almost flat, can cover a larger surface in a single blow than the typical slipper or boot. Sometimes, a poorly created hairbrush will break when it comes in close contact with an especially dangerous boy however, generally it's an extremely efficient remedy.

"Mrs. McFarren from Bristol, R.I., has a young boy who is now 6 yearsold, who has caused her great anxiety. [...] One year ago Mrs. McFarren was prevailed upon to purchase an electric hairbrush in the hope of improving the health of her hair. Since the brush was unusually massive and robust one, she decided to use it to educate her son. The first time the brush was used on him, it was because he had committed one of the most savage juvenile crimes and was consequently penalized with greater punishment. His mother was shocked when as soon as he was released the prisoner sprang into action and did several hand-springswhile in addition to breaking into songs. The rest of the day, he was in the best of mood, and not even an ounce of his old sadistic behavior was apparent. It was an utterly extraordinary and totally unprecedented situation that his mother was able to account for it solely on the assumption that the results from four decades of constant punishments had been cumulative and was only beginning to manifest itself.

"The boy was always getting in trouble and was, as a matter of fact regularly penalized. Every time the hairbrush was placed on him, his spirits seemed to increase and his activity level grew. Additionally, he started to gain weight and become stronger and all of his body ailments vanished. By the end of the year it was clear that he was the largest and heaviest young man that was his own age the entire town. And even though his inexplicably active lifestyle often caused him to break laws of his mother but nothing could stop the flow of his spirit or derail his unending great humor.

"There is no doubt that these amazing modifications to the physical and mental constitution of the McFarren little boy were the result of its electrical characteristics of the hairbrush that his mother used during the last year. The electric current, which was infused into his body by the force of the impact of the brush, brought him optimism and provided an impetus to the growth of his physical body. It is evident that the power of the electric hairbrush to deliver electricity to the scalp and therefore stimulate the growth of hair is the least valued feature. It will then be used not just as an instrument for juvenile punishment however, but as the most effective and safest method to inject vitality to the weak and weak of any age, as well as the elderly and weak, as well. Mrs. McFarren's name will remain forever linked with the most important of electrical discoveries."

The demand for electric household items lasted for a number of years, but the gas companies weren't eventually successful in convincing the general public to reject electricity as a risk. Edison Electric received between three and four thousand requests from private clients to set up electrical lighting in prisons, hotels, and factories the period from 1881 until 1882, the majority of which Edison Electric was forced to reject. Anyone who needed electricity at their location in the absence of street in which Edison's men were setting up the power grid, was required to construct their own generators, which was much more costly and difficult for Edison's business than it was financially profitable.

The Pearl Street Station is a success Pearl Street Station

While he could have made an enormous amount of money from the hundreds of customers interested in partnering in partnership with Edison Electric to construct their own power plants on their premises, Edison viewed such projects as an obstacle to his main goal. He was focused to a future where each American city was connected to the grid of power, and every person who desired electricity could access it by connecting to an existing power grid. The creation of grids across the country was essential element in making electricity an integral and essential part to American life.

(Edison was much more inclined to construct generators and lay in electric power for clients in other countries, as it was not necessary to spend so much money on an infrastructure.) The first step to achieve this objective was of course to electrify Manhattan however it took longer than expected to complete the first section of the network. As the delay grew, gas company stock began to increase and this was a sign that public's trust in Edison began to decline. His reputation suffered a further hit when he constructed an electric generator at the house of billionaire William Vanderbilt, only to be forced to demolish it at the direction of Vanderbilt's wife. This was after the newly installed electrical wire caused the wallpaper of the living room to begin to smoke.

Unscathed by the fire at the Vanderbilts house and a fellow billionaire J. Pierpont Morgan, the biggest privately held investor in Edison Electric, had a generator plant constructed on his property and electric lighting (including an incredibly electric desk lamp portable which was not ever before seen) were installed in his house. The desk lamp prototype that was wired to the desk's structure using a plate of metal in the ground, led to the whole desk to burn the moment it was turned on. However, like the Mrs. Vanderbilt, Morgan merely demanded that the device be rebuilt in a way that was less likely to ignite.

Edison Electric completed work on the underground conductors on Pearl Street on September 4 1882. The company then began providing electric lighting service to around 300 customers. The effect of lighting indoors disappointed certain people. The experience of electric lights was restricted to the arc lights that illuminated the streets and arc lighting was hundred times better than incandescent bulbs Edison employed. When the lights turned on in The New York Times and the Herald The writers and editors who had been working by the light of gas jets and candles extolled the excellence of Edison's electrical lighting in the most warm terms. However, in general the public did not pay much attention to this significant time in the technological history: Edison had been promising to illuminate Manhattan since 1878. four years after electricity was not a new phenomenon. According to Stross says, "the moment [was] an uplifting conclusion of a long-running story that included indelivered promises and a loud chorus of skeptical voices."

Edison offered his first group of clients four months free service, while he developed an instrument that could count the amount of electric power used each month. This was required to determine the amount to charge. Customers were thrilled enough to avail this deal initially however, when electricity started to be expensive it was noticed that there was a widespread resistance to switching from gas, despite the fact that Edison claimed it would cost more affordable. At the end of the day, only 231 customers remained with Edison Electric's service after their trial ended. This was a distant from the huge commercial performance Edison had hoped for. At the close of one year of operation, number customers had grown to 455, which was far from the numbers that Edison had anticipated.

Edison had made it known that he planned to build an additional power station in a different area of Manhattan following the opening of the one located at Pearl Street, but this proved to be unpractical in the short run because Pearl Street did not begin to earn profits until it had been operating over two years (and at that point, it was not until Edison hired an outside manager and promised him a payout of 10,000 dollars if he managed to get Pearl Street out of red.) Edison's investors advised Edison to concentrate his efforts on the one product that he was able to sell at a rapid rate which was off-site generators. Although he was still believing that the future was on centralized electricity for every urban area however, he became resigned to the fact that distribution power plant was the sole Edison product which was then generating an annual profit. Edison was always unhappy the fact that he had to send his most skilled employees to cities and states far away to oversee the construction of distributed generators whenever they were needed in Manhattan to help Peal Street Peal Street project go more efficiently. But , with working on the Manhattan grid on the verge of a break that would last until the year 1888 and he let the distributed generators to be the full-time work of the majority people who worked for him.

Conclusion

Thomas Alva Edison was known as an independent thinker. He was inspired through Thomas Paine's The Age of Reason which was handed to Edison from his dad Samuel Edison. Edison was adamant about Paine's scientific deism, saying that "He was called an atheist, however wasn't him," (Israel 2000).).

A lot of Edison critics claimed that the man was an atheist. The idea was sparked around 1910 after a reporter Edward Marshall, interviewed Edison on the subject of the religion of his day. Marshall declared that Edison opposed the notion of the supernatural, mortality, the soul, and also a personal God. "Nature," said Edison is not loving or merciful however, it is completely merciless and uninterested (Vernon 2014). A famous psychologist remarked to Edison that "People who don't think that immortality is a myth are insane or even pathologic" (Vernon 2014).

The truth is Edison thought that the religion of God should put emphasis on moral and not theological topics. Churches should transform into real schools of ethics. They must not teach fables, and instead place the appropriate emphasis of what is known as the Golden Rule. Edison believed, however, in "Supreme Intelligence" so he was an orthodox 19th-century deist.

Edison was a believer in peace and nonviolence. He was asked as a naval advisor in World War I, but the contract stipulated that he could only focus with defensive weaponry. The nonviolent philosophy he embraced extended to the world of animals and his name was known for electrocuting a variety of cats and dogs and an elephant using direct and alternating current technology.

Edison was a man with a fascinating humorous sense. In 1920, he caused the media when he revealed to his readers in the American Magazine that he was developing a sprit telephone which would enable contact with dead. Newspapers and magazines enjoyed an era of their own with the announcements and Edison received visits, letters and even conversations regarding his belief. Finally he said, "I really had nothing to tell him, but I hated to disappoint him so I thought up this story about communicating with spirits, but it was all a joke," (Paranormal-Encyclopedia, 2001).

Edison was a gentleman of the globe. He was a defender of economic reforms within the United States and was opposed to the gold standard and debt-based money. He was quoted as saying that gold was an relic of Julius Caesar and interest is the work of Satan. It is possible to read about his views regarding money in his plan that was published in 1922, a proposed amendment to the Federal Reserve Banking System. He provided an explanation of a commodity-backed currency. His ideas were intriguing but they were lacked basis.

Edison was appointed an Honorable Consultant Engineer for the Louisiana Purchase Exposition World's Fair in 1904. Edison was awarded the American Association of Engineering Societies John Fritz Medal in 1908. Edison was awarded membership in the National Academy of Sciences in 1927. In 1928, Edison received the Congressional Gold Medal.

Edison continues to be rewarded and in 2008 was admitted into the New Jersey Hall of Fame. In 2012, Edison was awarded the Technical Grammy award, and in 2011 , he was recognized as a member of the Entrepreneur Walk of Fame.

Edison is regarded as one of the most successful businessmen in America. He played a key role in the growth of America its economy in the country's most vulnerable beginnings and the shift from an agricultural-based economy to one of industrialization. Whatever your opinion regarding Edison is, it's impossible to place him in the sidelines of history. Edison was America's most famous inventor with 1,093 patents under his name. It is the only one that has been undiscovered in the history of other innovators. He was the prototype for the modern industrial research.

In the years that followed his death, Edison kept thinking about the potential of electricity, energy and the best way to promote inventions. Edison was also interested in the atomic energy field and once stated, "There will one day come in the mind of Science an instrument or force that is so terrifying in its possibilities, and so terrifying that even the one who dares to murder and torture in order to inflict death and torture is going to be horrified and abandon combat for good," (Vernon, 2014). The truth is that these words have never come in the form of a reality.

www.ingramcontent.com/pod-product-compliance
Lightning Source LLC
Chambersburg PA
CBHW050025130526
44590CB00042B/1911